"十三五"国家重点出版物出版规划项目·航天先进技术研究与应用系列

Theories and Applications of Moving Particle Semi-implicit Method

移动粒子半隐式方法理论及应用

张天纲 著

哈尔滨工业大学出版社

内容简介

This book systematically introduces the moving particle semi-implicit (MPS) method and its improvements in the mechanical engineering. The book is composed of six chapters. The first chapter introduces the classification of mesh and meshless methods. The second chapter carefully describes the derivation of the MPS method. Chapter three reviews the different wall boundary conditions used in the MPS method and emphasizes the polygon wall boundary condition that is an important wall boundary condition widely employed in the industry simulations. Chapter four to six introduces the latest enhancements of the polygon wall boundary conditions in the particle number density and the pressure distributions.

The book has clear structure and pays attention to the basic theories and derivations. Many newest research progress of the MPS method are included. The book can be used as the textbook for the postgraduate students in the field of the mechanical engineering, nuclear engineering, naval architecture, ocean engineering and other relative majors. The researchers and engineers can also study the methods in the book to conduct industry simulations.

图书在版编目(CIP)数据

移动粒子半隐式方法理论及应用＝Theories and Applications of Moving Particle Semi-implicit Method/张天纲著. —哈尔滨:哈尔滨工业大学出版社,2019.8
ISBN 978－7－5603－7420－8

Ⅰ.①移… Ⅱ.①张… Ⅲ.①粒子模型 Ⅳ.①O571.21

中国版本图书馆 CIP 数据核字(2018)第 128111 号

策划编辑	张　荣
责任编辑	李长波
出版发行	哈尔滨工业大学出版社
社　　址	哈尔滨市南岗区复华四道街 10 号　邮编 150006
传　　真	0451－86414749
网　　址	http://hitpress.hit.edu.cn
印　　刷	黑龙江艺德印刷有限责任公司
开　　本	787mm×1092mm　1/16　印张 7.25　字数 230 千字
版　　次	2019 年 8 月第 1 版　2019 年 8 月第 1 次印刷
书　　号	ISBN 978－7－5603－7420－8
定　　价	48.00 元

(如因印装质量问题影响阅读,我社负责调换)

Preface

The computational fluid dynamics (CFD) is an important technique in the fluid dynamics in the 21th century. Employing the discretized mathematical methods to simulate the experiments of the fluid dynamics in the computers, the practical problems can be predicted and solved. With the enhancement of the ability of the computer calculation, the CFD has become an indispensable tool for the fluid simulations in recent years.

The CFD can be classified into two large categories, i. e., the mesh-based and mesh-free simulation methods. The moving particle semi-implicit (MPS) method is a Lagrangian approach belonging to the mesh-free method. The advantage of the MPS method is that it can readily track the fragmentation, coalescence and large deformation of the fluids, which is a difficulty in the mesh-based methods. In addition, the programming of the MPS method is simpler than mesh-based methods. Thus, the MPS method has been widely used in the fields of the numerical simulations especially in the industry simulations since it was invented. However, the book systematically introducing the MPS method is written in Japanese and there is no books domestically teaching the MPS method. In addition, the improved methods of the MPS method are scattered in many literatures, which is also difficult for the readers to study.

In view of these problems, the author writes this book for the researchers, engineers, postgraduate students and the readers who are interest in the numerical simulations, to satisfy their demands to understand this method. Because different people view numerical methods differently and to make the contents more self-contained, we include one chapter on the difference between the grid-based method and mesh-free method. Then the MPS method is systematically introduced. The wall boundary conditions used in the MPS method are elaborated because they are the main factor affecting the numerical accuracy. The other three chapters mainly introduce the improvements of the MPS method in recent years so that the readers can comprehend the high accurate methods employed in the industry simulations.

The author learned the MPS method from the professor Koshizuka, the inventor of the MPS method for three and a half years. Thus, the author is very familiar with the details of the MPS method. The contents of the book are all summarized and supplemented from the researches of the author. The methods introduced in this book have been mostly adopted by the industry companies or simulation software. The Hitachi Company supported the researches of the author and provided the CAD file of the complex geometry for the book. Without their support, this work would never have come into existence. Thus, I want to show my thanks to them.

Tiangang Zhang
2018. 11

Contents

Chapter 1 Introduction 1
  1.1 Computational fluid dynamics 1
    1.1.1 Grid methods 1
    1.1.2 Mesh-free methods 3
  1.2 Polygon wall boundary condition 3

Chapter 2 MPS method 4
  2.1 Governing equations 4
  2.2 Weight function and discretization models 4
  2.3 Derivation of MPS 6
  2.4 Wall boundary conditions 7

Chapter 3 Boundary conditions in MPS 9
  3.1 Five types of wall boundary conditions 9
    3.1.1 Dummy particles 9
    3.1.2 Mirror particles 14
    3.1.3 Boundary forces 17
    3.1.4 Unified semi-analytical wall boundary condition 18
    3.1.5 Polygon wall boundary condition 20
  3.2 Initialization of polygon wall boundary condition 22
  3.3 Discretization models of polygon wall boundary condition 23
    3.3.1 Derivation process 23
    3.3.2 Discretization equations 24

Chapter 4 Improved wall calculation of polygon wall boundary condition 28
  4.1 Problems of polygon wall boundary condition 28
  4.2 Improvement of wall calculations 32
    4.2.1 Illustration of improved method 32
    4.2.2 Calculation of the wall weight function 32
  4.3 Numerical examples 34
    4.3.1 Accuracy of wall weight function at different positions 34
    4.3.2 Classic dam break simulation 38
    4.3.3 Dam break simulation with a wedge in the water tank 42
  4.4 Summary 46

Chapter 5 Boundary particle arrangement technique in polygon wall boundary condition
.... 47

5.1 Boundary particle arrangement technique ··· 47
    5.1.1 Construction of the boundary particles ··································· 48
    5.1.2 Construction of dummy particles ·· 51
5.2 Adjustment of collision coefficients ·· 52
5.3 Simulation results ·· 52
    5.3.1 Dam break with a wedge ··· 52
    5.3.2 Simulation of complex geometry ··· 63
5.4 Summary ··· 69

Chapter 6 Improvement of pressure distribution in polygon wall boundary condition ·· 70

6.1 Research progress ··· 70
6.2 Improved particle-polygonal meshes interaction models ···················· 72
    6.2.1 Re-derivation of polygon wall boundary condition ················ 72
    6.2.2 Problem of present source term in the polygon wall boundary condition ··· 73
    6.2.3 Improvement of source term in the polygon wall boundary condition ··· 73
    6.2.4 Improvement of gradient model ·· 76
    6.2.5 Surface detection ·· 77
6.3 Results and discussions ·· 78
    6.3.1 Hydrostatic pressure ··· 78
    6.3.2 Dam break simulation ·· 82
    6.3.3 Complex geometry simulations ··· 92
6.4 Summary ··· 95

References ·· 97
Noun index ·· 107

Chapter 1 Introduction

1.1 Computational fluid dynamics

With the improvement of the performance of the computers, simulation becomes a more and more important tool for analyzing, designing, prototyping, testing, fabrication and operating complex systems without having to carry them out, which could save lots of money and time. Through proper simulation the hypotheses can be verified and the phenomena occurred in the complex systems can be clearly observed. Thus, simulation provides a way to comprehend the mechanism of the world around you.

1.1.1 Grid methods

The simulation model is the mathematics description of how the system works. In recent years the simulation models based on the partial differential equations have significant development. Generally, the frame of reference of a numerical method is divided into two types: Eulerian reference frame in which the coordinates of grid are fixed in space, not to the material, which moves across the grid or mesh cells, and the Lagrangian reference frame in which the coordinates of mesh or grid are fixed to the material along the simulation. The advantage of the Eulerian frame reference is that arbitrary large deformations can be easily accommodated. However, the history of a material parcel that is a small region of fluid is difficult to track. On the contrary, the Lagrangian reference frame can easily track the material history since the coordinates follow the deformations but the accuracy often relates to the geometry of the points. A comparison between Eulerian and Lagrangian methods[1] is introduced by Liu & Liu[2] in Table 1-1.

In the field of numerical simulation methods such as the finite difference method (FDM)[3-5], finite element method (FEM)[6,7], and the finite volume method (FVM)[8,9] are the popular numerical simulation methods which are defined on the meshes of data points. In these methods, each vertex has a fixed number of neighboring vertexes, and the connectivity between neighboring vertexes can be used to define the mathematical operators. The operators are then used to simulate the Euler or Navier-Stokes equations.

Table 1-1 A comparison between Eulerian and Lagrangian methods[1]

	Lagrangian methods	Eulerian methods
Grid	Attached on the moving material	Fixed in the space
Track	Movement of any point on materials	Mass, momentum and energy flux across grid nodes and mesh cell boundary
Time history	Easy to obtain time-history at a point attached on material	Difficult to obtain time-history data at a point attached on materials
Moving boundary and interface	Easy to track	Difficult to track
Irregular geometry	Easy to model	Difficult to model with good accuracy
Large deformation	Difficult to handle	Easy to handle

The FDM based on the Taylor expansion was firstly proposed to approximate the differential equations. The FDM uses the square grid to construct the discretization of the partial differential equations. However, it is difficult treat complex geometries in multiple dimensions. To solve the problem of complex geometries in multiply dimensions, the integral form of the partial differential equations was adopted and subsequently the FEM and FVM were developed.

FEM is based on the Lagrangian reference frame, and proposed to obtain an approximate solution for the complex boundary value problems. To obtain the approximate solution, the computational region is divided into many small connected subdomains, which is called finite element, and the approximate polynomial functions at the subdomains are used to represent the unknown quantity. Combining these small finite elements, the governing equation for the whole structure can be obtained.

FVM is represented by Eulerian reference frame. In FVM method, the Navier-Stokes equations over all the control volumes in the solution domain are integrated. A variety of finite difference approximation equations in the integrated equation representing flow processes are then applied. Thus, it converts the integral equations into a system of algebraic equations[8].

The errors in the grid based numerical methods are either caused by the numerical diffusion of material properties in Eulerian reference frame or by the interpolation errors of deformed Lagrangian grids. Since Eulerian and Lagrangian reference frames both have defects, some mixed methods coupled Eulerian with Lagrangian frame were developed. These methods include the arbitrary Lagrangian-Eulerian (ALE)[10] method, the particle in cell[11, 12] method, marker and cell[13] method and the material point method (MPM)[14].

1.1.2 Mesh-free methods

Contrasting to the grid based methods are the mesh free methods. The mesh free methods are under the Lagrangian reference frame. Thus, the governing equations are solved on a set of particles rather than on the background mesh. The advantage of the mesh free methods is that the numerical diffusion accompanying with the discretization of the convection term can be avoided. In addition, the large deformation and fragmentation of fluids can be easily treated. The mesh free methods discretize the governing equations in a simple way and they are easier to be implemented than grid based methods. In recent year, mesh free methods have been well developed and several classic methods have been presented such as the finite point method (FPM)[15-18], diffuse element method (DEM)[19], element free Galerkin (EFG)[20-23] method, reproduced kernel particle method (RKPM)[24-27], HP-cloud method[28-30], meshless local Petrov-Galerkin (MLPG)[31-34] method, point interpolation method (PIM)[35-37], smoothed particle hydrodynamics (SPH)[38, 39] method and moving particle semi-implicit (MPS)[40] method.

1.2 Polygon wall boundary condition

In the numerical simulation methods, boundary conditions play an important part. Appropriate boundary condition can enhance the simulation results. In MPS[40] method, a polygon wall boundary condition was presented by Harada et al.[41] to treat complex geometry. In this boundary condition, CAD data can be directly utilized without special treatment. To simplify the calculation of the wall boundaries, in the polygon wall boundary condition the information of the wall boundaries is transformed into the distance values which are stored at the underlying grid points. Thus, geometries can be simulated with high efficiency. In this way complex geometries in three-dimension(3-D) can be easily simulated. Since the advantages of simulating complex geometries with high efficiency, the polygon wall boundary condition becomes a very effective simulation method in industry simulations.

Although the polygon boundary condition is very effective in the simulation of complex geometry, the inaccurate wall boundary leads to unnatural motions of fluid particles and deteriorates the simulation.

Chapter 2   MPS method

As illustrated in Chapter 1, the polygon wall boundary condition was proposed based on the moving particle semi-implicit (MPS) method. Thus, in this chapter MPS method is first introduced. MPS method presented by Koshizuka and Oka[40] in 1996 is a mesh free, Lagrangian numerical simulation technique. Since MPS method can easily treat free-surface flow, especially large deformation and fluid fragmentation[42, 43], it can be extended to many applications including ocean engineering[44], bubble rising[45], nuclear engineering[46], chemical engineering[47], and mechanical engineering[48].

2.1   Governing equations

The governing equations for incompressible viscous flows are the mass conservation and Navier-Stokes equations:

$$\frac{D\rho}{Dt} = 0 \tag{2-1}$$

$$\frac{D\boldsymbol{u}}{Dt} = -\frac{1}{\rho}\nabla P + \nu\nabla^2 \boldsymbol{u} + \boldsymbol{g} \tag{2-2}$$

where ρ is the density; t is the time; $\boldsymbol{u}$ is the velocity vector; P is the pressure; ν is the dynamic viscosity, and $\boldsymbol{g}$ is the acceleration due to gravity.

2.2   Weight function and discretization models

In MPS method, the density and pressure increase when particles come close to each other and vice versa. The interaction between particles uses the weight function

$$w(r_{ij}) = \begin{cases} \dfrac{r_e}{r_{ij}} - 1 & (0 < r_{ij} < r_e) \\ 0 & (r_e \leqslant r_{ij}) \end{cases} \tag{2-3}$$

where r_{ij} is the distance between particles i and j; and r_e is the effective radius. Fig. 2-1 shows the relationship between r_{ij} and $w(r_{ij})$. r_e is chosen as $2.1 r_{ij}$. The weight function has diverse forms as shown in Table 2-1 where the different weight functions are introduced in [49, 50]. All the forms of the weight function satisfy that the weight values are nonzero in the region of effective radius and the weight function increases as r_{ij} decreases and vice versa.

The variation of the density is reflected on the variation of the particle number density. The particle number density is calculated by the summation of the weight function

$$n_i = \sum_{j \neq i} w(|\mathbf{r}_j - \mathbf{r}_i|) \qquad (2\text{-}4)$$

where $\mathbf{r}_i$ and $\mathbf{r}_j$ are the position vector of the ith and jth particles, respectively; and n_i is the particle number density of the ith particle.

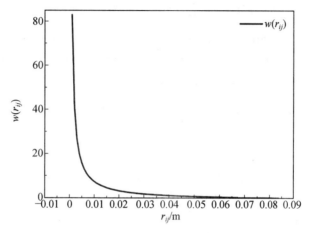

Fig. 2-1 Relationship between r_{ij} and $w(r_{ij})$

Table 2-1 Different weight functions

Weight functions	Reference
$w(r_{ij}) = \begin{cases} e^{-(r_{ij}/\alpha r_e)^2} & (0 \leqslant r_{ij} \leqslant r_e) \\ 0 & (r_e < r_{ij}) \end{cases}$	Belytschko et al. [51]
$w(r_{ij}) = \begin{cases} \dfrac{2}{3} - 4\left(\dfrac{r_{ij}}{r_e}\right)^2 + 4\left(\dfrac{r_{ij}}{r_e}\right)^3 & (0 \leqslant r_{ij} \leqslant \dfrac{r_e}{2}) \\ \dfrac{4}{3} - 4\left(\dfrac{r_{ij}}{r_e}\right) + 4\left(\dfrac{r_{ij}}{r_e}\right)^2 - \dfrac{4}{3}\left(\dfrac{r_{ij}}{r_e}\right)^3 & (\dfrac{r_e}{2} < r_{ij} \leqslant r_e) \\ 0 & (r_e < r_{ij}) \end{cases}$	Belytschko et al. [51]
$w(r_{ij}) = \begin{cases} 1 - 6\left(\dfrac{r_{ij}}{r_e}\right)^2 + 8\left(\dfrac{r_{ij}}{r_e}\right)^3 - 3\left(\dfrac{r_{ij}}{r_e}\right)^4 & (0 \leqslant r_{ij} \leqslant r_e) \\ 0 & (r_e < r_{ij}) \end{cases}$	Belytschko et al. [51]
$w(r_{ij}) = \begin{cases} -2\left(\dfrac{r_{ij}}{r_e}\right)^2 + 2 & (0 \leqslant \dfrac{r_{ij}}{r_e} < \dfrac{1}{2}r_e) \\ \left(2\dfrac{r_{ij}}{r_e} - 2\right)^2 & (\dfrac{1}{2}r_e \leqslant \dfrac{r_{ij}}{r_e} \leqslant r_e) \\ 0 & (r_e \leqslant r_{ij}) \end{cases}$	Koshizuka[52]
$w(r_{ij}) = \begin{cases} \dfrac{40}{7\pi r_e^2}\left[1 - 6\left(\dfrac{r_{ij}}{r_e}\right)^2 + 6\left(\dfrac{r_{ij}}{r_e}\right)^3\right] & (0 \leqslant r_{ij} < \dfrac{r_e}{2}) \\ \dfrac{10}{7\pi r_e^2}\left(2 - 2\dfrac{r_{ij}}{r_e}\right)^3 & (\dfrac{r_e}{2} < r_{ij} < r_e) \\ 0 & (r_{ij} > r_e) \end{cases}$	Shao and Lo[53]

The discretization models of MPS method are derived from the Taylor expansion. The gradient and Laplacian models in the MPS method are given as

$$\langle \nabla \phi \rangle_i = \frac{d}{n^0} \sum_{j \neq i} \left[\frac{\phi_j - \phi_i}{|r_j - r_i|^2} (r_j - r_i) w(|r_j - r_i|) \right] \quad (2\text{-}5)$$

$$\langle \nabla^2 \phi \rangle_i = \frac{2d}{\lambda n^0} \sum_{j \neq i} \left[(\phi_j - \phi_i) w(|r_j - r_i|) \right] \quad (2\text{-}6)$$

where ϕ_j and ϕ_i are quantities possessed by particles i and j, respectively; n^0 is the constant particle number density; d is the number of spatial dimensions; and λ is the Laplacian model coefficient, which is defined as

$$\lambda = \frac{\sum_{j \neq i} w(|r_j - r_i|) |r_j - r_i|^2}{\sum_{j \neq i} w(|r_j - r_i|)} \quad (2\text{-}7)$$

This parameter adjusts the increase in the variance caused by the Laplacian model to that of the analytical solution. The gradient and Laplacian models are substituted into their corresponding operators in Eq. (2-2) and the pressure, velocity and the position can be obtained. Substitute the velocity and pressure of each particle into the Lagrangian model. The viscosity term and the source term of pressure can be represented by

$$\langle \nabla^2 u \rangle_i = \frac{2d}{\lambda n^0} \sum_{j \neq i} \left[(u_j - u_i) w(|r_j - r_i|) \right] \quad (2\text{-}8)$$

$$\langle \nabla^2 P \rangle_i = \frac{2d}{\lambda n^0} \sum_{j \neq i} \left[(P_j - P_i) w(|r_j - r_i|) \right] \quad (2\text{-}9)$$

where u_i and u_j are the velocities of the ith and jth fluid particles, respectively; P_i and P_j are the pressure of the ith and jth fluid particles, respectively. Substitute the pressure of particles into the gradient model. The pressure gradient model can be obtained

$$\langle \nabla P \rangle_i = \frac{d}{n_0} \sum_{j \neq i} \left[\frac{P_j - \hat{P}_i}{|r_j - r_i|^2} (r_j - r_i) w(|r_j - r_i|) \right] \quad (2\text{-}10)$$

where $\hat{P}_i$ is the minimum pressure within the effective radius r_e. The purpose of replacing pressure P_i with $\hat{P}_i$ is to enforce repulsive force to all particles so as to stabilize the simulation.

2.3 Derivation of MPS

The semi-implicit scheme is adopted in MPS method. The Navier-Stokes equation is solved in two steps:

$$\left[\frac{Du}{Dt} \right]^{k+1} = -\left[\frac{1}{\rho} \nabla P \right]^{k+1} + \left[\nu \nabla^2 u \right]^k + \left[g \right]^k \quad (2\text{-}11)$$

The superscript k represents the kth time step, and the $k+1$ represents the $(k+1)$th time step. The left side of Eq. (2-11) can be divided into two parts

Chapter 2 MPS method

$$\left[\frac{D\boldsymbol{u}}{Dt}\right]^{k+1} = \frac{\boldsymbol{u}^{k+1} - \boldsymbol{u}^*}{\Delta t} + \frac{\boldsymbol{u}^* - \boldsymbol{u}^k}{\Delta t} \tag{2-12}$$

The explicit process is conducted. The intermediate velocity is only calculated by viscosity and gravity term

$$\boldsymbol{u}^* = \boldsymbol{u}^k + \Delta t \left[\nu \nabla^2 \boldsymbol{u}\right]^k + \Delta t \left[\boldsymbol{g}\right]^k \tag{2-13}$$

The intermediate position can be ascertained by

$$\boldsymbol{r}^* = \boldsymbol{r}^k + \boldsymbol{u}^* \Delta t \tag{2-14}$$

Then the incompressible condition is enforced in the MPS method; namely, the particle number density must remain constant, as

$$n_i = n^0 \tag{2-15}$$

where n_i is the particle number density of the ith fluid particle.

The following Poisson equation is derived, and the pressure of particles is implicitly evaluated using this equation

$$\langle \nabla^2 P \rangle_i^{k+1} = -\frac{\rho}{\Delta t^2} \frac{n_i^* - n^0}{n^0} \tag{2-16}$$

where n_i^* is the temporary particle number density calculated after the explicit step. After calculating the pressure of particles, the gradient of pressure is computed by Eq. (2-10). The velocity of next time step is corrected by

$$\boldsymbol{u}^{k+1} = \boldsymbol{u}^* - \Delta t \left[\frac{1}{\rho} \nabla P\right]^{k+1} \tag{2-17}$$

The position of particles at $(k+1)$th time step can be calculated

$$\boldsymbol{r}^{k+1} = \boldsymbol{r}^* + (\boldsymbol{u}^{k+1} - \boldsymbol{u}^*) \Delta t \tag{2-18}$$

The flow chart of the MPS method is shown in Fig. 2-2.

2.4 Wall boundary conditions

The Dirichlet boundary condition is enforced in the Poisson equation on the free surface. The pressure of particles is set to zero when the particle number density satisfies the equation

$$n_i^* < \beta n^0 \tag{2-19}$$

where β is a parameter for surface detection and $\beta = 0.97$ is adopted in this study.

The solid wall boundary condition is represented by fixed boundary particles whichare composed by one layer of boundary particles and several layers of dummy particles.

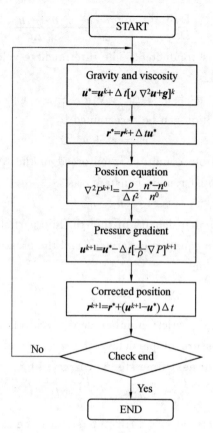

Fig. 2-2　Flow chart of the MPS method

Chapter 3   Boundary conditions in MPS

Since any simulation has region, boundary condition becomes an important factor that needs to be well treated in the simulations. The performance of the simulations is not only affected by the algorithm but also influenced by the boundary treatment. In this chapter, we will summarize and categorize all the wall boundary condition in particle methods and illustrate the advantage and disadvantage of each wall boundary condition. Then introduce the polygon wall boundary condition in detail.

3.1   Five types of wall boundary conditions

In particle methods, there are mainly five types of wall boundary conditions including the dummy particles, mirror particles, boundary forces, unified semi-analytical wall boundary condition and polygon wall boundary condition. In common, dummy particles and mirror particles are collectively called ghost particles. The five types of boundary conditions will be introduced respectively.

3.1.1   Dummy particles

Since the treatment to the wall boundary can become very complicate, MPS method adopts the boundary particles the same as fluid particles to represent the wall boundaries. These boundary particles are called dummy particles. Dummy particles use one layer of boundary particles to calculate the pressure and two or more layers of particles to supplement the weight function. The number of layers of boundary particles relates to the effective radius. Large effective radius needs more layers of dummy particles to supplement the deficiency of the particle number density. The arrangement of the boundary particles in MPS method is shown in Fig. 3-1 where the grey particles are the first layer of boundary particles that take part in the calculation of pressure and the other white particles are the dummy particles that are not involved in the pressure calculation. Fig. 3-2 shows the interaction between the fluid particles and the boundary particles. The black particles with grey circles particles in Fig. 3-2 are the fluid particles and r_e is equal to 2.1 times of diameter of fluid particles. The dummy particles can supplement the particle number density. Since the dummy particles have the same properties as the fluid particles and it is easy to be implemented, this boundary condition has been adopted by many simulations[54-56].

Shakibaeinia and Jin[50] also adopted the dummy particles proposed by Koshizuka and Oka[40] to simulate the inlet and outlet boundary conditions. To the solid wall boundary, several layers of dummy particles are arranged. The dummy particles are interpolated to the fluid region. The normal velocity of dummy particles is set to zero. For the free-slip boundary condition, the tangential velocity is equal to the fluid particle. For the no slip boundary condition, the tangential velocity is opposite to the fluid particle. Pressure is calculated at the boundary particles and extrapolated to other layers of dummy particles. In SPH[38, 39] method, Federico et al.[57] also employed the inflow and outflow boundary condition to simulate the free surface flow. The difference between these two methods is that Federico et al.[57] used the Froude number to restrict the outflow velocity of fluid particles. To improve the efficiency, a weakly compressible model[58-61] which has already been used in the smoothed particle hydrodynamics[38, 39] method is used to improve the efficiency.

Fig. 3-1 Arrangement of the boundary particles in MPS method

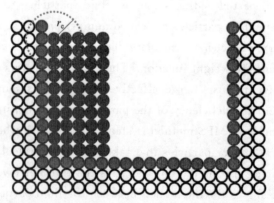

Fig. 3-2 Interaction between the fluid particles and the boundary particles

To the standard inlet and outlet boundary condition, the inflow particles must be equal to the outflow particles so as to keep the mass conservation. The Fig. 3-3 shows the sketch of standard inlet and outlet boundary conditions. The dummy particles are utilized at the inflow and outflow wall boundaries the same as MPS method to supplement the deficiency of the particle number density to void too small particle number density of fluid particles near the inflow and outflow boundaries. When the fluid particles flow out the outflow boundary, they are moved to the inflow boundary. However, to the unequal number of inflow and outflow particles, this technique cannot be utilized. Shakibaeinia and Jin[50] presented a recycling strategy. A type of special particles called "storage particles" is utilized in the inlet and outlet boundary condition. The storage process is shown in Fig. 3-4. The outflow particles are not directly moved to the inflow boundary, but stored at the storage region as storage particles. The physical properties of the fluid particles are removed from the fluid particles. The particles entering the inflow boundary are abstracted from the storage region and the physical properties are added to these particles. Thus, the recycling strategy can maintain the mass conservation.

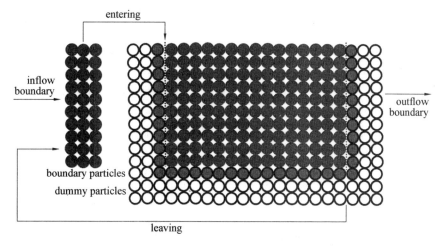

Fig. 3-3 Inlet and outlet boundary conditions

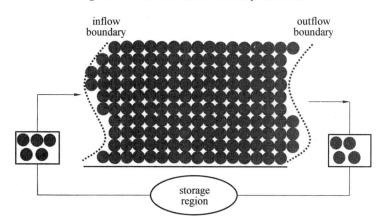

Fig. 3-4 Particle recycling strategy

To the inlet boundary in [50], several layers of dummy particles are defined at the inflow boundary to compensate the particle number density. To the velocity boundary condition, the fluid particles are added at depth y relating to the average distance Δl

$$k(y) = \frac{\Delta l}{u(y)\Delta t} \tag{3-1}$$

where $u(y)$ is the inflow velocity of the particle at depth y; Δt is the time step and $k(y)$ is a coefficient to designate the time when fluid particles is added to the inflow boundary. The particles are added to the inflow boundary according to the equation below

$$t = k(y)\Delta t \tag{3-2}$$

Thus, at different depth, the inflow particles can be added naturally.

To the outflow boundary, several layers of dummy particles are defined. The first layer of boundary particles that are close to the fluid particles are designated with pressure (depth) as shown in Fig. 3-3. When fluid particles approaching the dummy particles, they are removed and stored at the storage region.

The concept of dummy particles is also used in other particle methods. In SPH[38, 39] method, Morris et al.[62] utilized dummy particles to simulate a curved surface which is similar to the "imaginary particles" in [63] that obtain functional forms for the viscous force from the wall boundary. The several layers of dummy particles are arranged along the wall boundaries in [62]. The closest distance d_a from fluid particle a to each polygon is calculated. Then a tangent plane (a line in two dimension) is defined. The distance d_B from each dummy particle B in the effective radius to the tangent plane or line can be calculated as shown in Fig. 3-5. Then the velocity of dummy particle B is calculated by

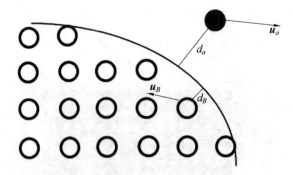

Fig. 3-5　Construction of velocity for boundary particles

$$\boldsymbol{u}_B = -(d_B/d_a)\boldsymbol{u}_a \tag{3-3}$$

No-slip boundary condition can be enforced. Jones et al.[64] proposed an empirical weighting correction to emulate curved surface when boundary particles clump together with dummy particles. Colagrossi et al.[65] also used the dummy particles to impose free-slip boundary condition to simulate two-dimensional interfacial flows.

Adami et al.[66] proposed a local force balance between the fluid particles and the wall boundary which is represented by layers of dummy particles to simulate two and three dimensional geometries in WCSPH[67-69]. The free-slip and no-slip boundary conditions can be easily enforced and good pressure distribution can be obtained.

Although dummy particles can be easily used to simulate three dimensional geometries, it is difficult to represent the geometry accurately in the complex geometries. To solve this problem, Sun et al.[70] developed a cylindrical coordinate system to improve the accuracy of the arrangement of boundary particles in agitator simulation. However, this method can only be applied to regular geometries or geometries that can be represented by curvilinear functions.

Since the treatment to the dummy particles is the same as that to the fluid particles and the dummy particles can be easily extended to two and three dimensional problems, this wall boundary condition has been applied to different type of simulations[54, 71].

In spite of the advantages of dummy particles, Li et al.[72] simulated freely falling fluid particles to the flat wall boundary represented by dummy particles. The trajectories of the fluid particles at different positions are different. To illustrate the problem of the wall boundary composed by the dummy particles, they use one fluid particle freely falling to the flat wall boundary. The Poisson's equation is represented by

$$[\sum_{j \neq i, j \in \text{wall}} w(|\mathbf{r}_j - \mathbf{r}_i|)] p_i - \sum_{j \neq i, j \in \text{wall}} w(|\mathbf{r}_j - \mathbf{r}_i|) p_j = \frac{\lambda \rho (n^* - n^0)}{2 d \Delta t^2} \quad (3-4)$$

When the fluid particle comes close to the wall boundary, the trajectory of the fluid particle is not natural because the particle number densities are different even at the same height to the wall boundary as depicted in Fig. 3-6 where fluid particle i has the same distance d to the dummy particles. The particle number density of fluid particle i in Fig. 3-6(b) is smaller than that in Fig. 3-6(a), which causes the non-uniform contribution of the wall boundary to the fluid particle. Through this example and other validations, Li et al.[72] verify that that a discretized numerical wall cannot accurately represent real physical wall boundaries. In addition, to the complex geometries, dummy particles cannot represent the geometry accurately as illustrated above. Thus, these deficiencies restrict the application of the dummy particles.

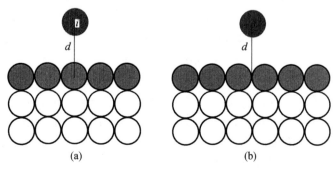

Fig. 3-6 Different positions of the fluid particles that have the same height to the wall boundary

3.1.2 Mirror particles

Besides the dummy particles, mirror particles[73, 74] are another widely used wall boundary condition. The mirror particles are the fluid particles near the wall boundary reflected across the wall boundary to supplement the particle number density.

Pozorski & Wawreńczuk[75] used the mirror particles in SPH method. The mirror particles possess the velocity, density and mass but not involved in the calculation of the dynamics. The position and velocity of the mirror particles are computed by

$$\boldsymbol{r}_{a'} = 2\boldsymbol{r}_w - \boldsymbol{r}_a \qquad (3\text{-}5)$$

$$\boldsymbol{u}_{a'} = 2\boldsymbol{u}_w - \boldsymbol{u}_a \qquad (3\text{-}6)$$

where a, a' and w represents the fluid particle, mirror particle and wall boundary, respectively. The no-slip boundary condition can be easily enforced. The relationship between fluid particle a and mirror particle a' is shown in Fig. 3-7 where no-slip boundary condition is enforced.

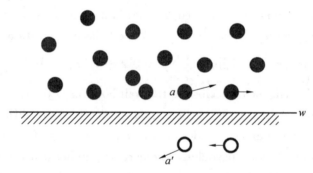

Fig. 3-7 The relationship between fluid particle a and mirror particle a'

Akimoto[76] simulated the flow around a planing body using mirror particles in MPS method. To the flat wall boundary, the mirroring procedure is the same as the method by Pozorski & Wawreńczuk[75]. The pressure is also considered in the method by Akimoto[76]. The normal vector is also introduced to designate the direction of the wall boundary segments. The velocity and position of the mirror particles are computed by

$$\boldsymbol{x}'_i = \boldsymbol{x} - 2l_s \boldsymbol{n} \qquad (3\text{-}7)$$

$$\boldsymbol{u}'_i = \begin{cases} \boldsymbol{u}_i - 2[(\boldsymbol{u}_i - \boldsymbol{u}_w) \cdot \boldsymbol{n}]\boldsymbol{n} & \text{(free-slip wall)} \\ 2\boldsymbol{u}_w - \boldsymbol{u}_i & \text{(no-slip wall)} \end{cases} \qquad (3\text{-}8)$$

where $\boldsymbol{x}$ and $\boldsymbol{x}'_i$ are the position vectors of fluid particle i and mirror particle, respectively; $\boldsymbol{n}$ is the normal vector of the wall boundary; $\boldsymbol{u}_i$, $\boldsymbol{u}'_i$, and $\boldsymbol{u}_w$ are the velocity vectors of the fluid particle i, mirror particle and the wall boundary, respectively; l_s is the distance from fluid particle i to the wall boundary.

The pressure of the mirror particle is also computed by

$$p'_i = p_i + (\boldsymbol{x}'_i - \boldsymbol{x}_i) \cdot \boldsymbol{K} \qquad (3\text{-}9)$$

where $\boldsymbol{K}$ is the body force; p_i and p'_i are the pressure of the fluid particle i and mirror particle, respectively.

When a fluid particle is around the corner the generally used mirror technique will lead to deficiency or excess of the particle number density, which will destroy the flow field. To address this issue, the convex and concave are discussed separately as shown in Fig. 3-8 where white particles are the mirror particles and black ones are the fluid particles. To the concave corners as shown in Fig. 3-8 (a), the corner is divided by equal apex angle and the contribution of the mirrored particles is ignored if the fluid particles are mirrored to the other side of the angular bisector as the dotted white particle in Fig. 3-8 (a). On the contrary, if the corner is convex as shown in Fig. 3-8 (b) the mirrored particles on the other side of the angular bisector are also computed. By this simple treatment, all the two-dimensional problems can be easily treated. Cherfils et al.[77] also used mirror particles to simulate free surface flows with parallel computing techniques.

Yildiz et al.[78] simulated a two-dimensional cylinder obstacle in a channel flow with an improved mirror particle method named multiple boundary tangent method. This method extended the tangent line approach of the method proposed by Morris et al.[62] to a curved surface and used a single layer of dummy particles to represent the wall boundary. At regular time intervals, a fluid particle is reflected across the wall boundary through the tangent line of every dummy particle inside the effect radius of the fluid particle. Overlapping specular particles are then deleted. The remaining specular particles are used to compute the particle number density. The multiple boundary tangent method can be extended to three dimensions. However, repetitive operations of mirroring and deleting particles are time-consuming, especially in three dimensions.

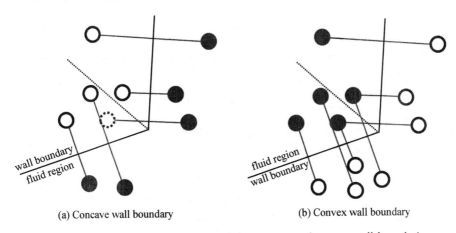

(a) Concave wall boundary (b) Convex wall boundary

Fig. 3-8 Mirroring process around the concave and convex wall boundaries

Marrone et al.[79] use the idea of the mirror particles to generate dummy particles. However, the dummy particles are only generated once. First, the wall boundary is represented by boundary particles with equal space d. Then, the boundary particles are duplicated the distance of $d/2$ outside of the geometry in the normal direction. A spline function is defined according to these duplicated boundary particles. Along the spline, the dummy particles are generated with equal space d. Then repeat this process, several layers of boundary particles are generated. The advantage of the fixed dummy particles is that distribution of dummy particles is always uniform and the generated dummy particles do not depend on the position of the fluid particles. After generating the dummy particles, all the dummy particles are mirrored back into the fluid field to form the interpolation points. If the wall boundary is right angle or flat surface, the interpolation points are directly mirrored by the dummy particles. However, to other angles, special treatments need to be done carefully to guarantee the proper positions of the interpolation points. The relationship between the dummy particle and the interpolation point is shown in Fig. 3-9. The free-slip and no-slip boundary condition can be imposed on the interpolation points. The Neumann boundary condition can be enforced, namely

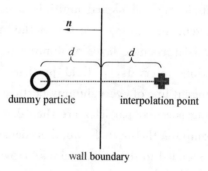

Fig. 3-9 The relationship between the dummy particle and the interpolation point

$$\frac{\partial p}{\partial n} = \rho \boldsymbol{f} \cdot \boldsymbol{n} \tag{3-10}$$

where $\boldsymbol{f}$ is the body force; $\boldsymbol{n}$ is the normal vector of the wall boundary. Substitute Eq. (3-10) into the SPH model and the pressure of dummy particles can be calculated

$$p_d = \sum_{j \in \text{fluid}} p_j W^{\text{MLS}}(\boldsymbol{r}_j) dV_j + 2d\rho \boldsymbol{f} \cdot \boldsymbol{n} \tag{3-11}$$

where W^{MLS} is the moving least-squares kernel which can be calculated by the method proposed by Fries & Matthies[80].

The mirror particles can accurately maintain the shape of the geometries and easily enforce the free-slip or no-slip boundary conditions. In two dimensional simulations, mirror particles method can obtain good results. However, mirror particles are difficult to be extended to three dimensions. Although a few methods can be extended to three dimensions, they can only treat simple shapes with low efficiency.

Since mirror particles can enforce accurate free-slip and no-slip boundaryconditions, which can supplement the deficiencies of the dummy particles, some mixed methods coupling the mirror particles with the dummy particles were developed. Park and Jeun[81] developed a coupled method with dummy and mirror particles. In this method, single layer of dummy particles are used to designate the wall boundary, and mirror particles are generated temporarily at each time step to improve the overall stability as show in Fig. 3-10.

Lee et al.[82] also utilized mixed boundary conditions with ghost and mirror particles. In their method, three layers of dummy particles are used to compute the wall contribution namely the wall weight function, and mirror particles are constrained by free-slip and no-slip boundary conditions at each time step. Although coupling methods can improve the accuracy of simulations, it is difficult to extend mirror particles to three dimensions. Thus, the application range of coupling methods is limited.

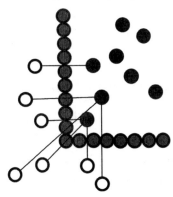

Fig. 3-10 Mirroring procedure near the corner

3.1.3 Boundary forces

To simplify the arrangement of wall boundary particles, a boundary force was proposed by Monaghan[59, 67, 83-85]. Unlike ghost particles, a boundary force requires a single layer of wall particles to represent the wall boundary. The repulsive force is enforced by the Lennard-Jones potential between the fluid and boundary particles

$$f(r) = D\left(\left(\frac{r_0}{r_{ij}}\right)^{p_1} - \left(\frac{r_0}{r_{ij}}\right)^{p_2}\right) \frac{r_{ij}}{r_{ij}^2} \tag{3-12}$$

where r_{ij} is the distance between fluid particle i and the wall boundary particle j; r_0 is the initial distance between particles; p_1 and p_2 are chosen as 4 and 2 or 12 and 6. Coefficient $D=5gH$ where H is the depth of the water. $D=gH$ or $D=10gH$ can also obtain similar results. This method can simulate complex geometries. However, de Leffe et al.[86] noted that the Lennard-Jones potential is suitable for straight wall boundaries. When representing non-planar wall boundaries, the Lennard-Jones potential suffers from spurious behavior. To address this issue, Monaghan et al.[87] proposed another method using boundary force integrals to suppress unphysical motions of fluid particles near wall boundaries. The force from the wall boundary can be calculated by

$$\sum_{j=-\infty}^{+\infty} \Gamma(jL_0) = \frac{1}{L_0}\int_{-\infty}^{+\infty}\Gamma(q)\mathrm{d}q + 2\sum_{n=1}^{+\infty}\int_{-\infty}^{+\infty}\Gamma(q)\cos(\frac{2\pi nq}{L_0})\mathrm{d}q \qquad (3\text{-}13)$$

where q is the distance between fluid particle and boundary particle; L_0 is the initial distance between particles. This method is more stable than that using the Lennard-Jones potential. However, the boundary force is represented by integral form which is lack of unified discretization equations. Thus the discretization equation is different according to each problem, which restricts the application range of this method.

In sum, the boundary forces can treat three dimensional problems even the complex geometries. The wall boundary is represented by single layer of boundary particles which is simpler than the dummy particles. The repulsive force is enforced by simple function without considering the shape of the geometry. However, the boundary forces are not accurate. The unphysical motion of the fluid particles occurs near the wall boundary.

3.1.4 Unified semi-analytical wall boundary condition

Another boundary condition with a single layer of boundary particles is the unified semi-analytical wall boundary condition used in SPH method. This boundary condition naturally combines the boundary integral with particle interaction models. Macià et al.[88] defined the Shepard normalization factor

$$\gamma_h(x) := \int_a^b W_h(x-y)\mathrm{d}y \qquad (3\text{-}14)$$

where W_h is the weight function. The SPH approximation with respect to W_h is defined as

$$\langle p \rangle(x) := \frac{1}{\gamma_h(x)}\int_a^b p(y)W_h(x-y)\mathrm{d}y \qquad (3\text{-}15)$$

where $p(y)$ is a scalar function. In one dimension, assume discretization points $\{x_1, x_2, \ldots, x_N\}$ equally spaced at interval (a, b). Based on the Eq. (3-15) the discretization of the first-order derivative can be derived

$$\langle \frac{\mathrm{d}p}{\mathrm{d}x} \rangle_i = \frac{1}{\gamma_i}\sum_j \frac{m_j}{\rho_j} p_j W'_h(x_i-x_j) + \frac{1}{\gamma_i}[p_b W_h(b-x_i) - p_a W_h(a-x_i)] \qquad (3\text{-}16)$$

The discretization of the second-order derivative can also be obtained by

$$\langle \frac{\mathrm{d}^2 p}{\mathrm{d}x^2} \rangle_i = \frac{2}{\gamma_i}\Big[\sum_j \frac{m_j}{\rho_j}\frac{p_i-p_j}{x_i-x_j}W'_h(x_i-x_j)$$
$$+ \frac{p_i-p_b}{x_i-x_b}W_h(b-x_i) - \frac{p_i-p_a}{x_i-x_a}W_h(a-x_i)\Big] \qquad (3\text{-}17)$$

where p_j and m_j are the density and the mass of particle j, respectively. By this way, the boundary terms are naturally derived in the discretization equations. However, these formulations can only be used in one dimension. In higher dimensional problems, the boundary terms will become very complicated.

To treat the problem of higher dimensions, Ferrand et al.[89] proposed an unified semi-analytical wall boundary conditions and derived a new incompressible SPH (ISPH)[74, 90, 91] model using boundary integrals. The idea of this boundary condition is the same as the method by Macià et al.[86]. The boundary integral is also defined on the computational region

$$\gamma_a := \int_{\Omega \cap \Omega_a} w(r' - r_a) \mathrm{d}V' \tag{3-18}$$

where Ω and Ω_a represent the computational region and the region of fluid particle a respect to the wall boundary Ω, respectively; w is the weight function; r_a is the position vector of particle a. If a has no truncation with the wall boundary, γ_a is equal to 1. Otherwise, γ_a is smaller than 1. The relationship between fluid particle a and wall boundary is shown in Fig. 3-11. Then the density is estimated by

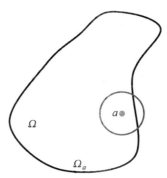

Fig. 3-11 The relationship between fluid particle a and wall boundary

$$\rho_a = \frac{1}{\gamma_a} \sum_{b \in \text{fluid}} m_b w_{ab} \tag{3-19}$$

where γ_a is a variable representing the wall boundary. With γ_a, the discretization equations contain the wall boundary automatically. To accurately compute the contribution of the wall boundaries, the boundary integral is employed. The contribution of each boundary segment (in two dimensions) is calculated respectively as shown in Fig. 3-12 where every segment is composed by the vertexes, angle and normal vector. The derivative of γ_a can also be represented by the integral of the segments in the influence radius of a

$$\nabla \gamma_{as} = \left(\int_{r_{e_1}}^{r_{e_2}} w(r) \mathrm{d}l_s \right) \boldsymbol{n}_s \tag{3-20}$$

This boundary condition can obtain very good pressure distribution in two dimensions as applied by Leroy et al.[92]. However, in three dimensional simulations[93], the boundary integrals are very time consuming even in simple three dimensional geometries. To improve the efficiency, the boundary integrals are replaced by approximate equations, which dramatically reduce the accuracy. Thus, this method is

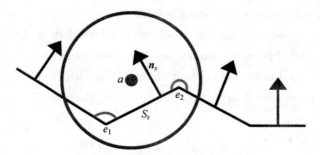

Fig. 3-12 Definition of boundary property

difficult to use for complex geometries in three dimensions. In sum, unified semi-analytical wall boundary condition still cannot treat complex geometries in three dimensions. Additionally, the same problem occurred as the ghost particles, i. e. , accurately arranging even a single layer of boundary particles along the wall boundary is difficult if the geometry is irregular or the physical shapes of the geometry do not satisfy certain functions.

To other method with single layer of boundary particles, Mayrhofer et al. [94] simulated rigid bodies using a single layer of boundary particles with a consistent renormalized scheme. However, this method does not consider the free surface.

In brief, unified semi-analytical wall boundary condition is mainly used in the two dimensional simulations. In three dimensions, this wall boundary condition can only treat simple geometries. Thus, the dummy particles, boundary forces and the unified semi-analytical wall boundary condition all have the problem that cannot accurately represent the boundary of the complex geometries except the mirror particles.

3.1.5 Polygon wall boundary condition

Polygonal wall boundary condition is another boundary condition that can accurately represent the wall boundaries. As for the treatment of polygonal walls, Kuasegaram et al. [95] derived the contact forces with variational formulation. Based on the work of Kuasegaram et al. [95], Feldman and Bonet[96] improved the wall weight function near angles and corners in two dimensions. The polygon wall boundary is split by straight wall boundary segments near corners. The contribution of each segment is calculated independently. Li et al. [72] proposed a new wall boundary condition represented by polygons to avoid non-uniform wall contributions. In this method, the wall contributions are also calculated from the contributions of the boundary segments, as in the methods by Kuasegaram et al. [95] and Feldman and Bonet[96]. However, linear combinations of boundary segments can only treat simple geometries in two dimensions. In addition, the contributions of boundary segments must be accurately computed to avoid inaccurately representing wall contributions. Thus, this method is simply an attempt to employ polygons to replace wall boundary particles.

In the MPS method[40], the boundary condition represented by polygons proposed by Harada et al.[41] can treat complex geometries and has been widely used in industry simulations as shown in Fig. 3-13. The wall boundaries of the geometry in the polygon boundary condition are not composed by particles but by polygons. The merits of this boundary condition are that the wall weight function can be calculated in a simple manner without using boundary particles and complex geometries can be simply treated. However, the low accuracy near non-planar polygons severely affects the simulation results. Watanabe et al.[97] used the same idea as Harada et al.[41] in SPH method and calculated the wall weight function with the curvature of the triangle patches proposed by Meyer et al.[98] and volume amendments. However, this method still suffers from the inaccurate particle number density of the wall boundary condition. Yamada et al.[99] proposed an explicit polygon boundary condition based on the MPS method with the same idea as the method by Harada et al.[41]. This method suffers from the strong pressure oscillation near the corners and angles. To suppress the unphysical motions of fluid particles near the polygons, Mitsume et al.[100] developed a new explicit polygon boundary condition which calculates the closest distance with a search algorithm without using the distance function. Mirror particle and a boundary force are employed as the mixed boundary condition. Although this method is more accurate than the method by Yamada et al.[99], the boundary force causes the pressure oscillation and the calculation of the distance is also time consuming especially in complex geometries.

The polygon wall boundary condition is derived from the MPS method. To simulate complex geometries, the distance function is introduced into the boundary condition. Next we will introduce the polygon wall boundary condition in detail.

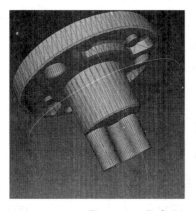

Fig. 3-13 Definition of boundary property

3.2 Initialization of polygon wall boundary condition

In the MPS method proposed by Koshizuka and Oka[40], the wall boundary is discretized by dummy particles, and the fluid and the wall particles are computed in the same manner. When the wall boundary is represented by polygons, the boundary particles disappear. To reinstate the contribution of the lost boundary particles, Harada et al.[41] proposed a method that divides the discretization equations of the MPS method into the wall and fluid parts. To represent the wall part, a uniform grid is introduced into this boundary condition before the main loop as shown in Fig. 3-14 where the geometry is represented by polygon and a uniform grid is utilized. At each grid point in the uniform grid, the closest distance from each grid point to each polygon, the normal vector, and the curvature are calculated and stored.

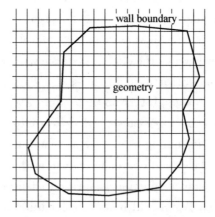

Fig. 3-14 Relationship between grid points and the geometry

The closest distance is computed from each grid point to each vertex, edge and surface of each polygon. The normal vector of each grid point $N_{ijk} = (N_i, N_j, N_k)$ is calculated by the central difference of the distance stored at the grid points

$$n_i = \frac{D_{i+1jk} - D_{i-1jk}}{2d} \tag{3-21}$$

$$n_j = \frac{D_{ij+1k} - D_{ij-1k}}{2d} \tag{3-22}$$

$$n_k = \frac{D_{ijk+1} - D_{ijk-1}}{2d} \tag{3-23}$$

where D_{ijk} is the distance value stored at grid point (i, j, k); d is the spacing between grid points. Normalize $\boldsymbol{n}_{ijk} = (n_i, n_j, n_k)$ and the normal vector can be calculated by

$$N_i = \frac{n_i}{|\boldsymbol{n}_{ijk}|} \tag{3-24}$$

$$N_j = \frac{n_j}{|\boldsymbol{n}_{ijk}|} \tag{3-25}$$

$$N_k = \frac{n_k}{|\boldsymbol{n}_{ijk}|} \tag{3-26}$$

The curvature is computed at each grid point i by the difference of normal vector

$$\kappa = \nabla^2 D(\boldsymbol{x}) = \nabla \cdot \boldsymbol{N} = \frac{N_{i+1} - N_{i-1}}{2d} + \frac{N_{j+1} - N_{j-1}}{2d} + \frac{N_{k+1} - N_{k-1}}{2d} \tag{3-27}$$

3.3 Discretization models of polygon wall boundary condition

3.3.1 Derivation process

The polygon boundary condition can be derived from the implicit process in MPS method. In the explicit process, the external force and viscosity are computed. The pressure caused by the external force and the viscosity pushes the fluid particles to the wall boundary. In the implicit process, the pressure is calculated. The wall boundary will push the fluid particles back to their balance positions, which is enforced by the calculation of the correction velocity in the implicit process

$$\boldsymbol{u}' = -\frac{\Delta t}{\rho} \nabla P^{k+1} \tag{3-28a}$$

Multiply Δt on both sides of Eq. (3-28a). The correction distance vector can be obtained

$$d\boldsymbol{r} = -\frac{\Delta t^2}{\rho} \nabla P^{k+1} \tag{3-28b}$$

The gradient term between fluid particle i and the wall boundary can be directly represented by

$$\langle \nabla P \rangle_{iw} = \frac{d}{n_0} \frac{P_w - P_i}{|\boldsymbol{r}_{iw}|^2} \boldsymbol{r}_{iw} Z_i \tag{3-29}$$

where P_w is the pressure of the wall boundary and Z_i represents the wall weight function. $\boldsymbol{r}_{iw} = \boldsymbol{r}_w - \boldsymbol{r}_i$ which means the closest distance from fluid particle i to each polygon. The relationship between the fluid particle i and the wall boundary is shown in Fig. 3-15 where L_0 is the initial distance between particles. $d\boldsymbol{r}$ and $\boldsymbol{r}_w$ satisfy

$$\frac{d\boldsymbol{r}}{|d\boldsymbol{r}|} = -\frac{\boldsymbol{r}_{iw}}{|\boldsymbol{r}_{iw}|} \tag{3-30}$$

Thus, Eq. (3-28) can be transformed to

$$P_w - P_i = \frac{\rho n_0 |d\boldsymbol{r}| |\boldsymbol{r}_{iw}|}{d \Delta t^2 Z_i} \tag{3-31}$$

where d is the dimensional number. To the 2-D problems, d is 2; To the 3-D problems,

d is equal to 3. Eq. (3-31) is the core concept of the polygon wall boundary condition.

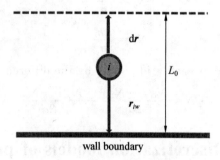

Fig. 3-15 The relationship between the fluid particle i and the wall boundary

3.3.2 Discretization equations

Using Eq. (3-31) the interaction between fluid particle and the wall boundary can be clearly represented. Thus, divide all the discretization equations of MPS method into the fluid and wall parts and substitute Eq. (3-31) into the wall part. The viscosity, source term and the gradient term of the wall part can be obtained, respectively.

Viscosity term

$$\langle \nabla^2 \boldsymbol{u} \rangle_i = \langle \nabla^2 \boldsymbol{u} \rangle_{if} + \langle \nabla^2 \boldsymbol{u} \rangle_{iw} \qquad (3\text{-}32)$$

where $\langle \nabla^2 \boldsymbol{u} \rangle_{if}$ uses Eq. (2-8) the same as MPS method. The wall part can be represented by

$$\langle \nabla^2 \boldsymbol{u} \rangle_{iw} = \frac{2d}{\lambda n^0} (\boldsymbol{u}_w - \boldsymbol{u}_i) Z_i \qquad (3\text{-}33)$$

where $\boldsymbol{u}_w$ is the velocity of the wall boundary. The wall boundary is assumed to have the same velocity.

Source term

$$\langle \nabla^2 P \rangle_i = \langle \nabla^2 P \rangle_{if} + \langle \nabla^2 P \rangle_{iw} \qquad (3\text{-}34)$$

where $\langle \nabla^2 P \rangle_{if}$ uses Eq. (2-9) the same as MPS method. The wall part can be represented by

$$\langle \nabla^2 P \rangle_{iw} = \frac{2\rho |d\boldsymbol{r}| |\boldsymbol{r}_{iw}|}{\lambda \Delta t^2} \qquad (3\text{-}35)$$

Gradient term

$$\langle \nabla P \rangle_i = \langle \nabla P \rangle_{if} + \langle \nabla P \rangle_{iw} \qquad (3\text{-}36)$$

where $\langle \nabla P \rangle_{if}$ uses Eq. (2-10) the same as MPS method. The wall part can be represented by

$$\langle \nabla P \rangle_{iw} = \frac{\rho |d\boldsymbol{r}| \boldsymbol{r}_{iw}}{\Delta t^2 |\boldsymbol{r}_{iw}|} \qquad (3\text{-}37)$$

Directly calculating r_{iw} is very time consuming. Thus, r_{iw} is interpolated from the closest distances stored at the grid points around fluid particle i.

The particle number density is also divided into two parts

$$n_i = \sum_{j \in \text{fluid}, j \neq i} w(r_{ij}) + Z_i \tag{3-38}$$

The fluid part is calculated by $\sum_{j \in \text{fluid}, j \neq i} w(r_{ij})$, the same as Eq. (2-3), the MPS method. To the wall part, the flat and non-planar wall boundaries are calculated respectively. Generally, $w(r_{iw})$ is used to represent the wall weight function of the flat wall boundary. To compute the wall weight function of the flat wall boundary, assume layers of dummy particles are arranged below the flat surface as shown in Fig. 3-16. To each distance r_{iw}, the wall weight function can be calculated. Thus, a set of correspondence between r_{iw} and $w(r_{iw})$ can be obtained. The relationship between r_{iw} and the wall weight function $w(r_{iw})$ is shown in Fig. 3-17. The wall weight function of each fluid particle can be linearly interpolated from Fig. 3-17 according to the closest distance of fluid particle.

To the non-planar wall boundary, the wall weight function Z_i is calculated by the degree of curve of the flat surface. The curvature is used to represent the degree of curve. Using curvature, the bending angle can be calculated by equation below

$$\kappa_i L_0 = \frac{L_0}{r_e} = 2\cos\theta_i \tag{3-39}$$

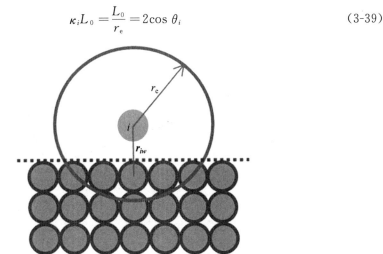

Fig. 3-16 Sketch of the position between fluid particle and dummy particles

where κ_i is the curvature of fluid particle i interpolated by the curvatures stored at the grid points. Fig. 3-18 shows the relationship between angle θ_i and the curvature. Using Eq. (3-39), angle θ_i can be obtained

$$\theta_i = \arccos \frac{\kappa_i L_0}{2} \tag{3-40}$$

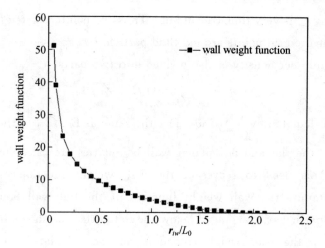

Fig. 3-17　The relationship between the wall weight function and the distance to the flat surface

The wall weight function of non-planar wall boundary can be calculated according to the change of angle

$$Z_i = \frac{\theta_i}{\pi} w(r_{iw}) \tag{3-41}$$

If the wall boundary is flat namely $\theta_i = \pi$, the wall weight function of i is equal to that of the flat wall boundary. In SPH method, the curvature is also used to calculate the wall contribution such as [99, 100]. The flow chart of the polygon boundary condition is shown in Fig. 3-19.

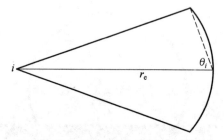

Fig. 3-18　The relationship between angle θ_i and the curvature

Chapter 3 Boundary conditions in MPS

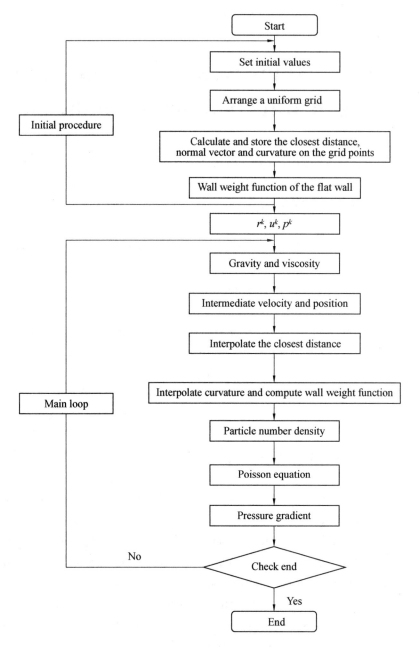

Fig. 3-19 The flow chart of the polygon boundary condition

Chapter 4 Improved wall calculation of polygon wall boundary condition

In this section, we clarify the reason for the low accuracy of the wall weight function proposed by Harada et al. [41] and propose a new formulation to improve the wall weight function near the non-planar wall boundary. The grid points outside the boundary are used to compute the wall weight function of the non-planar wall boundary. The ratio of the wall weight function by the non-planar wall boundary to that of the flat surface is stored at the grid points. The wall weight function of the fluid particles can be obtained through the trilinear interpolation of the values stored at the grid points around it.

We verify our method of comparing the wall weight functions against the method by Harada et al. [41] through hydrostatic simulation. The classic dam break problem is tested using our proposed method and the method by Harada et al. [41] to compare the obtained distributions of the particle number density. Finally, a dam break with a wedge is simulated to verify the applicability of our method to complex geometries.

4.1 Problems of polygon wall boundary condition

The wall boundary condition by Harada et al. [41] inaccurately renders the particle number density near the non-planar wall boundary. The particle number density contributed by the neighboring fluid particles, which is given by the first term of the right-hand side of Eq. (3-38), is the same as that in the original MPS method. Thus, the wall weight function Z_i is the main reason for the inaccuracy.

The wall weight function is calculated in three steps: the calculations of $|\bm{r}_{iw}|$, $w(\bm{r}_{iw})$, and Z_i. We compare the wall weight function by Harada et al. [41] and that of the original MPS method at different positions in the water tank as depicted in Fig. 4-1. The length, width, and height of the water tank are 1.24, 0.68, and 0.68 m, respectively. We test a uniform grid with grid spacings of $0.1L_0$, $0.5L_0$, and L_0, where the diameter L_0 of each fluid particle is 0.04 m. The effective radius r_e is $2.1L_0$. Point A is above the center of the bottom plane. Point B is along the line bisecting the angle between the left and bottom planes. Point C is along the line bisecting the angle formed where the three planes of the tank walls, the bottom, left, and front planes, intersect. First, we calculate the wall weight function of point A at different heights from the bottom plane. In this case, $Z_i = w(r_{iw})$ because $\theta_i = \pi$. The results are shown in Fig. 4-2. The wall weight function of point A is the same as that of the original MPS method despite the grid spacing. This illustrates that $|r_{iw}|$, $w(r_{iw})$, and Z_i are accurate near the flat wall boundary.

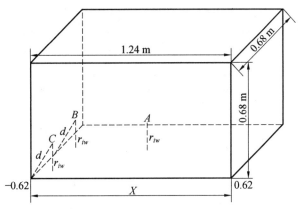

Fig. 4-1 Schematic description of the positions of particle i and the wall boundary

Next, we consider point B, which is a distance d from the edge of the tank, as shown in Fig. 4-1. Fig. 4-3 shows the relationship between $|r_{iw}|$ and d for different grid spacings. As the grid spacing decreases, the relation between d and $|r_{iw}|$ gradually changes to a linear relation because $|r_{iw}|$ is evaluated by the interpolation at the neighboring grid points. Thus, $|r_{iw}|$ is affected by the grid spacing near the non-planar wall boundary. Actually, it is expected that d and $|r_{iw}|$ have a linear relation, indicating that a smaller grid spacing is more accurate. However, the computation effort increases as the grid spacing decreases. The same situation occurs at point C, as shown in Fig. 4-4.

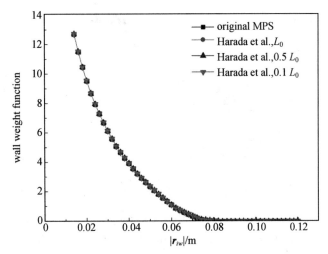

Fig. 4-2 The relationship between wall weight function and $|r_{iw}|$ at point A

Next, the third step is tested. We assume that the first step is accurate; namely, $|\boldsymbol{r}_{iw}|=\frac{\sqrt{2}}{2}d$. The second step of obtaining $w(\boldsymbol{r}_{iw})$ is always accurate. In Fig. 4-5, the wall weight functions by Harada et al. [41] and the original MPS method are represented by grey and black dots, respectively. The substantial difference between the black and grey dots in Fig. 4-5 is caused by the curvature, which is represented by θ_i in Eq. (3-41). Thus, the main problem in the wall boundary condition by Harada et al. [41] is the inaccuracy of the curvature near the non-planar boundary.

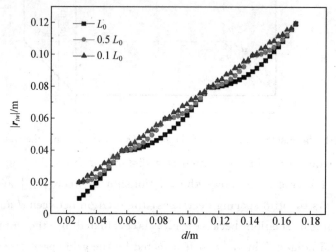

Fig. 4-3 The relationship between $|\boldsymbol{r}_{iw}|$ and d at point B

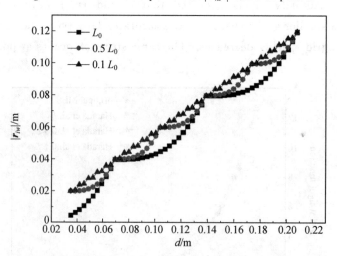

Fig. 4-4 The relationship between $|\boldsymbol{r}_{iw}|$ and d at point C

Finally, we compare the wall weight function of the method by Harada et al. [41] with different grid spacings at point B in Fig. 4-1 with that of the original MPS method to investigate the influence of the grid spacing on the wall weight function. The results are shown in Fig. 4-6. The labels "Harada et al., L_0" "Harada et al., $0.5L_0$" and

"Harada et al. , 0.1L_0 " indicate wall weight functions with grid spacings of L_0, 0.5L_0, and 0.1L_0, respectively. The graph shows that the wall weight function at point B decreases when the grid spacing is decreased. As shown in Fig. 4-3, $|\boldsymbol{r}_{iw}|$ is interpolated more accurately when a smaller grid spacing is used. This means that the convergent curve with respect to the grid spacing using the method by Harada et al.[41] is much smaller than the curve using the original MPS method, as shown in Fig. 4-5. Thus, the curve with a spacing of 0.1L_0 in the method by Harada et al.[41] shows more deviation from the original MPS than that with a grid spacing of L_0. Because the distribution of the wall weight functions with different grid spacings at point C is similar to that at point B, we do not show the wall weight functions at point C.

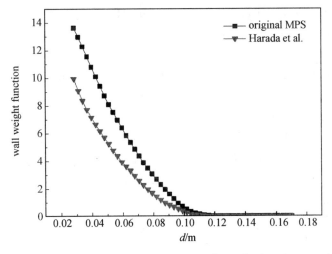

Fig. 4-5 The relationship between d and wall weight function at point B

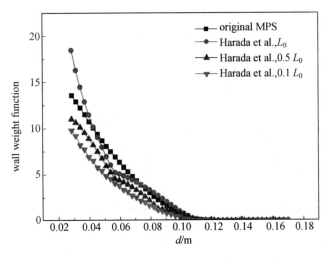

Fig. 4-6 Comparison of wall weight functions with different grid spacing at point B

4.2 Improvement of wall calculations

The inaccuracy of the wall weight function by Harada et al.[41] is derived from θ_i, which is calculated from the curvature in Eq. (3-40). Thus, replacing θ_i with a more accurate formulation is necessary.

4.2.1 Illustration of improved method

As explained above, the curvature in Eq. (3-41) is not an accurate representation of a non-planar wall boundary. We need to return to the idea of the particle wall boundary. Because an uniform grid is used in the method by Harada et al.[41], we regard the grid points outside the walls of the polygon as wall particles, and the wall weight function of the grid points inside or on the wall boundaries of the polygon is calculated at the grid points outside the boundary. A flow chart of the proposed method is given in Fig. 4-7. In the initial procedure of the proposed method, the closest distance and the normal vector are computed and stored in advance. The curvature does not need to be calculated; instead, a new coefficient is computed and stored at the grid points inside or on the polygon wall boundary. In the main loop, the wall weight function of the fluid particles near the polygon wall boundary is calculated based on the closest distance and the coefficient stored at the grid points.

4.2.2 Calculation of the wall weight function

In the initial procedure, the wall weight function is calculated at the grid points inside or on the polygon wall boundary as

$$Z_m = \sum_n w(r_{mn}, r_e) \tag{4-1}$$

where m is a grid point that is inside or on the polygon wall boundary and n is a grid point outside the polygon wall boundary. The right-hand side of Eq. (4-1) is calculated with Eq. (2-3).

After obtaining the wall weight, the coefficient C_m is calculated and stored as

$$C_m = \frac{Z_m}{W(r_{mw})} \tag{4-2}$$

where $W(r_{mw})$ is the wall weight function of the flat wall boundary and r_{mw} is the distance from the grid point m to the flat wall boundary. In the case of the flat wall boundary, $C_m = 1$. In the main loop, the wall weight function of fluid particles is calculated in three steps.

First, the distance is calculated from the linear interpolation of the distances stored at the surrounding grid points. Next, the coefficient C_i at fluid particle i is calculated from the linear interpolation of the coefficients stored at the surrounding grid points.

Finally, the wall weight function at fluid particle i is calculated by
$$Z_i = C_i w(\boldsymbol{r}_{iw}) \qquad (4\text{-}3)$$

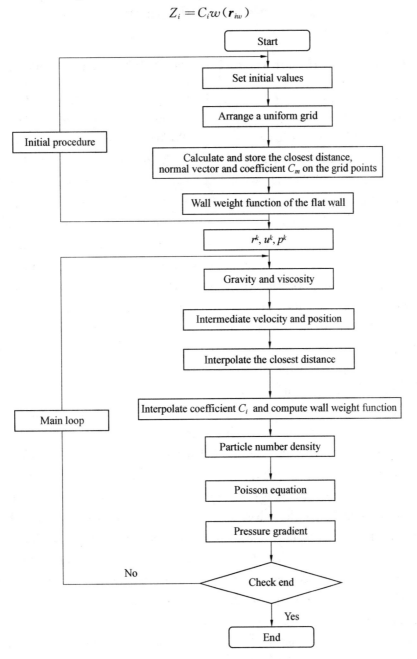

Fig. 4-7 A flow chart of the proposed method

4.3 Numerical examples

4.3.1 Accuracy of wall weight function at different positions

To evaluate the present wall boundary condition, we compare the method by Harada et al. [41], the method proposed in this study, and the original MPS method. The tank shown in Fig. 4-1 is considered here. We first compare the weight functions at point A with a grid spacing of L_0. Fig. 4-8 shows the results. The method by Harada et al. [41] and the present wall boundary condition exhibit good agreement with the original MPS method. Next, we compare the wall weight function near the non-planar wall boundary. Figs. 4-9 and 4-10 show the results at points B and C with grid spacing of L_0. The present method shows better agreement with the original MPS method than the method by Harada et al. [41]. The non-smooth curves from the present method are caused by the grid spacing. Because the wall weight function is calculated by the linear interpolation of the grid points, interpolation errors arise. The wall weight function of the method by Harada et al. [41] near the non-planar wall boundary is extraordinarily large when d is less than 0.04 m at point B and 0.05 m at point C. This problem produces an extraordinarily large pressure on the fluid particles. Thus, the simulation deteriorates near the non-planar wall boundary.

Next, we compare the wall weight functions of the improved wall boundary condition with different grid spacing. The referenced figure is also Fig. 4-1. The wall weight functions at points B and C with grid spacing of L_0, $0.5L_0$, and $0.1L_0$ are shown in Figs. 4-11 and 4-12, respectively. The accuracy of the wall weight function increases as the grid spacing decreases. However, in consideration of the computational effort, we use a grid spacing of L_0 in the subsequent simulations. Next, we compare the particle number density in the hydrostatic simulation. The rectangular tank in Fig. 4-1 retains water with a depth of 0.48 m. The bottom of the tank is 1.24 m ×0.68 m. Fig. 4-13 shows the distribution of the particle number density obtained using the method by Harada et al. [41] and the proposed method. The color of the particles represents the particle number density. The grid spacing L_0 is 0.04 m and the effective radius r_e is $2.1L_0$. The particle number density of the fluid particles from the method by Harada et al. [41] in Fig. 4-13(a) is smaller than n^0 near the non-planar wall boundary. The improved wall boundary condition in Fig. 4-13(b) shows a good distribution of the particle number density, even near the non-planar wall.

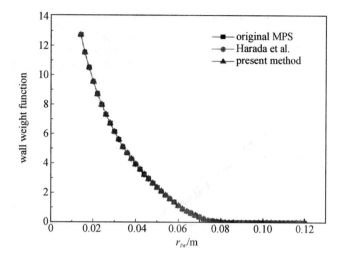

Fig. 4-8 Comparison of wall weight functions obtained using different methods at point A ($L_0 = 0.04$ m)

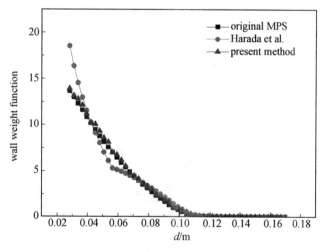

Fig. 4-9 Comparison of wall weight functions obtained using different methods at point B ($L_0 = 0.04$ m)

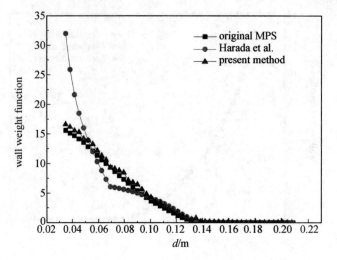

Fig. 4-10 Comparison of wall weight functions obtained using different methods at point C ($L_0 = 0.04$ m)

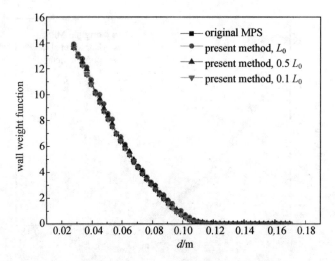

Fig. 4-11 Comparison of wall weight functions obtained using different methods at point B

Chapter 4 Improved wall calculation of polygon wall boundary condition

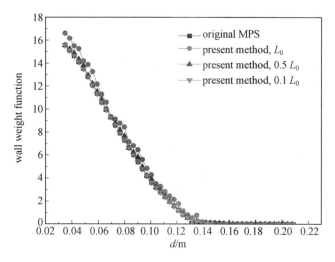

Fig. 4-12 Comparison of wall weight functions obtained using different methods at point C

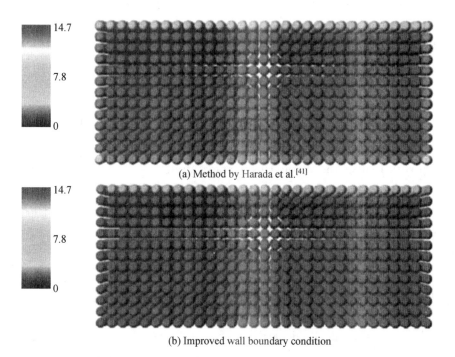

Fig. 4-13 Snapshots of the distribution of particle number density obtained by the method by Harada et al.[41] and the proposed method for hydrostatic simulation

4.3.2 Classic dam break simulation

In this section, we verify the improved wall boundary condition in the dam break simulation. The water tank is the same as that in Section 4.3.1. The length, height, and width of the water column are 0.52, 0.48, and 0.64 m, respectively. The diameter L_0 of the fluid particles is 0.04 m, and the effective radius r_e is $2.1L_0$. The number of fluid particles is 3 094, and the grid spacing is L_0.

Fig. 4-14 shows the inner ($n_i \geqslant 0.97n^0$) and surface ($n_i < 0.97n^0$) particles at $t = 0.0$ s. In the method by Harada et al.[41], the particles near the corners are judged as surface particles. However, in the proposed method, the particles near the corners remain inner particles. We then compare the particle number density near the wall boundary. Figs. 4-15 and 4-16 show snapshots of the dam break simulation at $t = 0.5$ and 1.0 s, respectively. The color represents the particle number density. The particle number density near the non-planar wall boundary fluctuates remarkably, and the fluid particles scatter above the free surface when the method by Harada et al.[41] is used. Conversely, in the proposed method, the fluid particles are much more stable, and unphysical motion does not appear. Figs. 4-17 and 4-18 show the distribution of particle number density of fluid particles near the bottom wall boundary. The black lines in Figs. 4-17 and 4-18 are the theoretical value n^0 of the particle number density, and the grey points are the particle number density of the fluid particles that are at a distance of $0.5L_0$ or less from the bottom surface at different x-coordinates, as shown in Fig. 4-1. The particle number density in the method by Harada et al.[41] is dispersed across a wide range, and some particles show values far smaller than the theoretical values. In comparison with that of the method by Harada et al.[41], the particle number density distribution of the fluid particles in the proposed method is close to the theoretical values. The reason for the scattered distribution of the particle number density in the method by Harada et al.[41] is the inaccurate representation of the wall weight function near the non-planar wall boundary.

More dots are visible in Fig. 4-17(b) than in Fig. 4-17(a). The reason for this is that the inaccurate wall weight function calculated using the curvature in the method by Harada et al.[41] causes a large fluctuation in the pressure field near the corners. This causes the velocities of the fluid particles near the corners to oscillate. On the other hand, in the proposed method, the particles are stable near the corners. Thus, the number of particles near the polygon wall in the method by Harada et al.[41] decreased.

Finally, the memory costs and CPU times of the method by Harada et al.[41] and the Proposed method are compared. The memory costs for the two methods are equal because the numbers of variables are the same.

The CPU times of the method by Harada et al.[41] and the Proposed method are

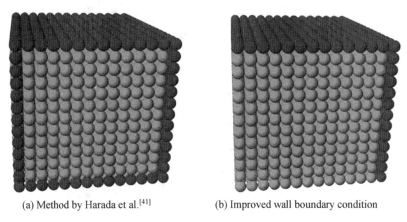

(a) Method by Harada et al.[41]   (b) Improved wall boundary condition

Fig. 4-14   Inner (grey) and surface (black) particles

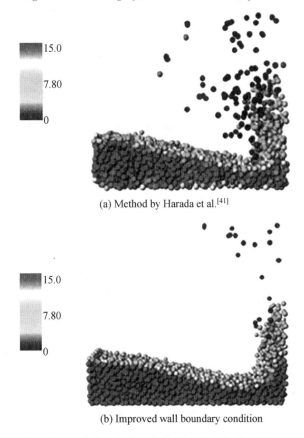

(a) Method by Harada et al.[41]

(b) Improved wall boundary condition

Fig. 4-15   Snapshots of the results of the dam break simulation at $t = 0.5$ s

compared in Table 4-1. The total CPU times of the method by Harada et al.[41] and the improved wall weight condition are 587.809 and 590.919 s, respectively, which are the sums of the CPU times of the initial procedures and the 2 000 reiterations of the main loops. The grid calculation is the CPU time required for the calculation of the uniform grid in the initial procedures of the two methods. The wall weight function is the CPU

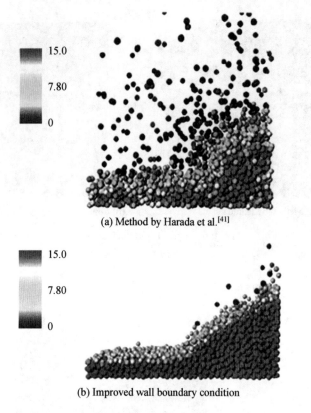

Fig. 4-16 Snapshots of the results of the dam break simulation at $t = 1.0$ s

time required for the specific calculations of the closest distance and the wall weight function in the main loop.

From Table 4-1, the CPU time required for the calculation of the wall weight function in the initial procedure of the new method is almost four times that of the method by Harada et al.[41]. In the main loop, the CPU time required to calculate the closest distance and the wall weight function in the new method is almost equal to that in the method by Harada et al.[41]. As a whole, the CPU times are almost the same for both methods. The specific calculations of the grid and the wall weight function in the method by Harada et al.[41] and the new method are 6.8% and 6.3% of the total computation time, respectively. Thus, the additional CPU time for the polygon wall boundary condition is small. Moreover, we expect that the CPU time would be substantially reduced by removing the wall particles, which are necessary in the original MPS method.

Chapter 4 Improved wall calculation of polygon wall boundary condition

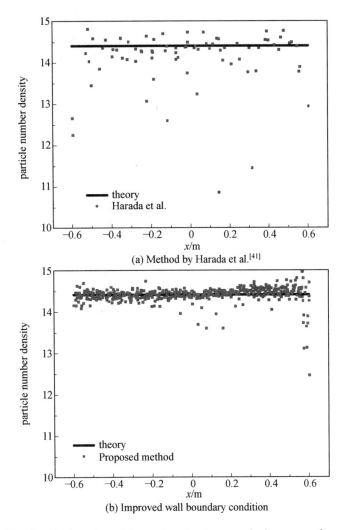

Fig. 4-17 The distribution of particle number density near the bottom surface at $t = 0.5$ s

Table 4-1 **CPU time required for dam break simulation using 2 000 main loops**

	initial procedure	grid calculation	main loop	wall weight function
Harada et al.	0.562 s	0.331 s	587.247 s	39.63 s
Proposed method	2.106 s	2.018 s	588.813 s	35.12 s

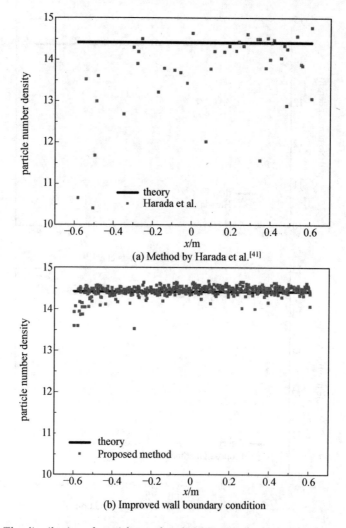

Fig. 4-18 The distribution of particle number density near the bottom surface at $t = 1.0$ s

4.3.3 Dam break simulation with a wedge in the water tank

In the previous examples, simple geometries with right angles were used. In this section, we simulate a water tank with a wedge that has a slope and an edge. The water tank geometry is depicted in Fig. 4-19. The length, width, and height of the water tank are 1.24, 0.68, and 0.96 m, respectively. There is a wedge inside the water tank. The left and right planes of the wedge are a slope and a plane perpendicular to the bottom plane of the water tank. The grid spacing L_0 and the effective radius r_e are the same as in Section 4.3.2.

Snapshots of the simulation results at $t = 1.22$ and 1.94 s are shown in Figs. 4-20 and 4-21. Color in the figures represents the particle number density. The distributions of the particle number density of the fluid particles near the slope (at a distance of less

Chapter 4　Improved wall calculation of polygon wall boundary condition · 43 ·

than or equal to $0.5L_0$) at $t = 1.22$ and $t = 1.94$ s are shown in Figs. 4-22 and 4-23, respectively. The black lines in Figs. 4-22 and 4-23 represent the theoretical values. The x-coordinate in Figs. 4-22 and 4-23 is the reference range depicted in Fig. 4-19. The simulation results are similar to those in Section 4.2. The number of fluid particles in the method by Harada et al. [41] is also smaller than that in the proposed method in Figs. 4-22 and 4-23. The reason is the same as that given in Section 4.3.2. Although the distribution of the particle number density near the incline region is not very good in the proposed method, the proposed method still obtains better results than the method by Harada et al. [41].

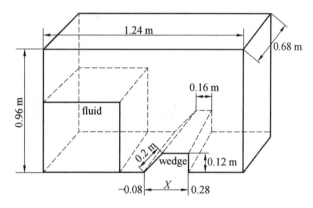

Fig. 4-19　Sketch of the simulated dam break against a wedge

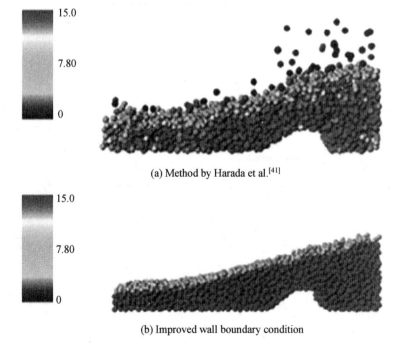

(a) Method by Harada et al.[41]

(b) Improved wall boundary condition

Fig. 4-20　Snapshots of the results of the simulated dam break against a wedge at $t = 1.22$ s

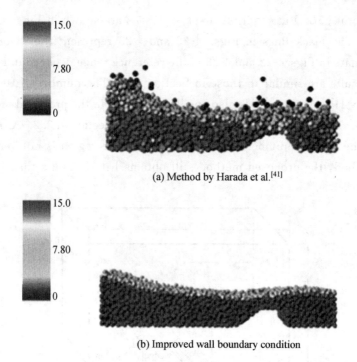

Fig. 4-21 Snapshots of the results of the simulated dam break against a wedge at $t = 1.94$ s

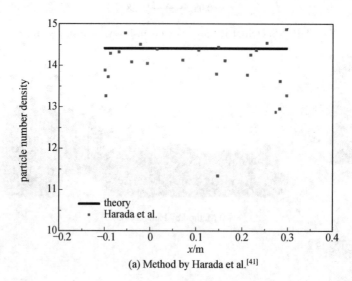

(a) Method by Harada et al.[41]

Fig. 4-22 Distribution of particle number density near the wedge at $t = 1.22$ s

Chapter 4　Improved wall calculation of polygon wall boundary condition

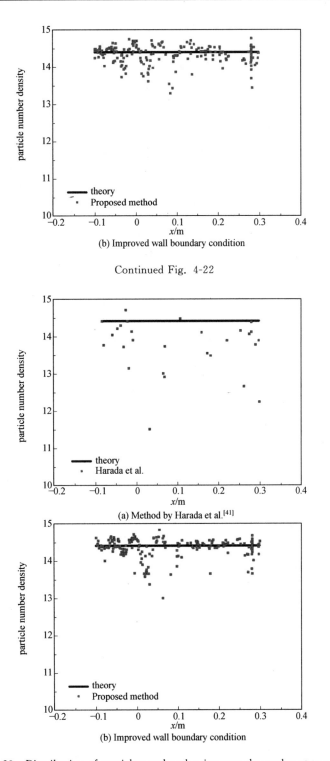

(b) Improved wall boundary condition

Continued Fig. 4-22

(a) Method by Harada et al.[41]

(b) Improved wall boundary condition

Fig. 4-23　Distribution of particle number density near the wedge at $t = 1.94$ s

4.4 Summary

In this section, the wall boundary condition proposed by Harada et al.[41] was analyzed carefully, and the reason for the inaccuracy of the wall weight function obtained using this method was clarified. This inaccuracy arises from the approximation of the non-planar wall boundary using the curvature. A new formulation was proposed to improve the wall weight function near a non-planar wall boundary. The grid points that are outside of the polygon wall are regarded as boundary particles. The wall weight function of the grid points inside or on the wall boundaries of the polygon is calculated at the grid points outside the boundary. A coefficient C_m is stored at the grid points inside or on the polygon wall. The wall weight function of the fluid particles can be calculated by the linear interpolation of the values stored at the nearby grid points. The improved boundary condition maintains the same level of simplicity and efficiency as the method by Harada et al.[41] Because the proposed method does not use the curvature, the unphysical motion of the fluid particles near the non-planar wall boundary is dramatically suppressed.

We calculated the wall weight function with a hydrostatic simulation, and the results demonstrate that the improved wall boundary condition is in good agreement with the original MPS, even near the non-planar wall boundary. The dam break simulation shows that the particle number density of the proposed method near the bottom surface is more accurate and stable than that of the method by Harada et al.[41]. The additional CPU time required in the present method is 6.3% of the total CPU time in the dam break simulation. Actually, it is expected that the total CPU time of the present method is smaller than that of the original MPS, which requires that wall particles be considered in the main loop. The simulation of a dam break against a wedge shows that our method still obtains a better particle number density near the slope than the method by Harada et al.[41].

Chapter 5 Boundary particle arrangement technique in polygon wall boundary condition

The wall weight function method proposed in Chapter 4 can improve the wall weight function near non-planar wall boundaries. However, the distribution of the particle number density near slopes as shown in Figs. 4-22 and 4-23 still has a large difference from the n^0. To address this issue, in this section, we propose an initial boundary particle arrangement technique coupled with the wall weight function method in Chapter 4 to improve the particle number density near slopes and curved surfaces with boundary conditions represented by polygons in three dimensions. Two uniform grids are utilized in the proposed technique. The grid points in the first uniform grid are used to construct boundary particles, and the second uniform grid stores the same information as in the wall weight function method. The wall weight functions of the grid points in the second uniform grid are calculated by newly constructed boundary particles. The wall weight functions of the fluid particles are interpolated from the values stored on the grid points in the second uniform grid. Because boundary particles are located on the polygons, complex geometries can be accurately represented.

The performance of the wall weight function method proposed in Chapter 4 with the boundary particle arrangement technique is verified in comparison with the wall weight function method without boundary particle arrangement by investigating two example geometries. The simulations of a water tank with a wedge and a complex geometry show the general applicability of the boundary particle arrangement technique to complex geometries and demonstrate its improvement of the wall weight function near the slopes and curved surfaces.

5.1 Boundary particle arrangement technique

Here, we propose an initial boundary particle arrangement (BPA) technique, which can be coupled with the wall weight function method (PBC) in Chapter 4 to improve the wall weight function near non-planar polygons in three dimensions. In this method, two uniform grids are utilized simultaneously. The first uniform grid is used to generate boundary particles on the polygons. The grid points in the second uniform grid are used to calculate and store the normal vector, closest distance, and coefficient C_m in Eq. (4-2), where Z_m is computed by the mth grid point and the boundary particles. The positions of the two uniform grids are not related. However, for simplicity, we set the two uniform grids to coincide with each other.

The first grid is used to generate boundary particles. Let Ω, Γ, and Ψ represent the regions inside, on, and outside of the polygons, as shown in Fig. 5-1(a), where the background grid is the first uniform grid. We focus on a portion of the polygons, namely the square region in Fig. 5-1(b), to illustrate the boundary particle arrangement technique.

The first uniform grid uses a grid spacing that is equal to the diameter of the fluid particle. Dummy particles are generated at the grid points, as shown in Fig. 5-1(c). G_{ijk} is employed to label the dummy particle at grid point (i, j, k). For any dummy particle G_{ijk}, $G_{ijk} \in \Omega \cup \Gamma \cup \Psi$. The closest distance from dummy particle G_{ijk} to any given polygon can be calculated as shown in Fig. 5-1(d), where the black line segment is perpendicular to the foot and G'_{ijk} on the polygon. If a dummy particle is on a polygon, such as G_{i-1jk} in Fig. 5-1(d), the length of the perpendicular segment is zero, and G'_{i-1jk} is the same as G_{i-1jk}.

The boundary particles are located on the polygons by moving the dummy particles outside of and on the polygons. In the initial boundary particle arrangement technique, the boundary particles are located layer by layer. The number of layers is determined by $L = \max\{n \in \mathbb{Z} \mid n \leqslant r_e\}$. In the present study, all simulations employ $r_e = 3.1L_0$. Thus, three layers of boundary particles are set, meaning $L = 3$.

5.1.1 Construction of the boundary particles

The boundary particles in the first layer are located first. If a dummy particle is on a polygon, this particle is regarded as the first layer, i.e.,

$$\text{If } G_{ijk} \in \Gamma, \text{ then } G_{ijk} \in L_1$$

where L_1 represents the first layer of the boundary particles, as depicted in Fig. 5-1(e), in which the circle G'_{ijk} is a dummy particle on a polygon.

After ascertaining the positions of the dummy particles on the polygons, other boundary particles in the first layer are determined.

For $\forall G_{ijk} \in \Psi$, if $\exists p \in \{G_{i+1jk}, G_{i-1jk}, G_{ij+1k}, G_{ij-1k}, G_{ijk+1}, G_{ijk-1}\} \cap \Omega$, i.e., if one of the neighboring dummy particles of particle G_{ijk} is inside the polygon, then G_{ijk} is moved to the polygon as the boundary particle on the first layer:

$$\boldsymbol{G'}_{ijk} = \boldsymbol{G}_{ijk} + d\boldsymbol{n} \qquad (5-1)$$

where $\boldsymbol{G'}_{ijk}$ is the position vector on the polygon after $\boldsymbol{G}_{ijk}$ is moved, as shown in Fig. 5-1(f), where d is the distance between vectors $\boldsymbol{G}_{ijk}$ and $\boldsymbol{G'}_{ijk}$

$$d = |\boldsymbol{G'}_{ijk} - \boldsymbol{G}_{ijk}| \qquad (5-2)$$

The normal vector $\boldsymbol{n}$ can be obtained as

$$\boldsymbol{n} = \frac{\boldsymbol{G'}_{ijk} - \boldsymbol{G}_{ijk}}{|\boldsymbol{G'}_{ijk} - \boldsymbol{G}_{ijk}|} \qquad (5-3)$$

Chapter 5 Boundary particle arrangement technique in polygon wall boundary condition

After they have been moved, the boundary particles on the first layer are located as shown by the circles on the wall boundary in Fig. 5-1(g).

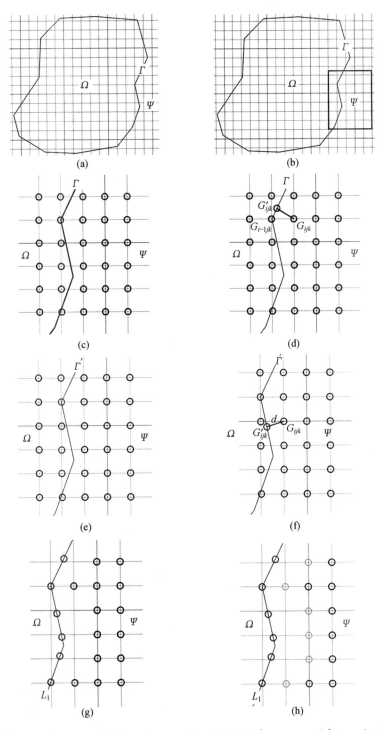

Fig. 5-1 Process of constructing boundary particles((a)-(q) show sequential steps in the process)

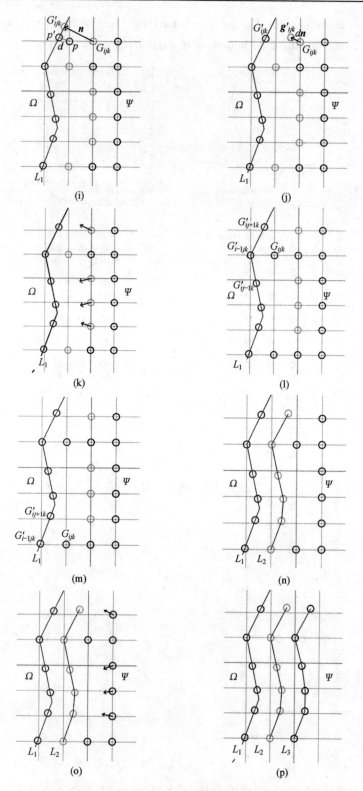

Continued Fig. 5-1

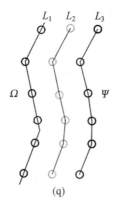

(q)

Continued Fig. 5-1

5.1.2 Construction of dummy particles

In this section, we describe the construction of the second layer L_2.

It is assumed that $G_{ijk} \in \Psi$ and particle G_{ijk} is not moved to the first layer. If one of the neighboring dummy particles of G_{ijk} is moved to L_1, G_{ijk} is moved to L_2, as shown by the green particles in Fig. 5-1(h):

$$\boldsymbol{g}'_{ijk} = \boldsymbol{G}_{ijk} + d\boldsymbol{n} \qquad (5-4)$$

where $\boldsymbol{g}'_{ijk}$ is the position vector on L_2 after G_{ijk} is moved. The normal vector $\boldsymbol{n}$ is calculated using Eq. (5-3). d is calculated as

$$d = |\boldsymbol{p}' - \boldsymbol{p}| \qquad (5-5)$$

where $\boldsymbol{p}$ is the vector of the neighboring dummy particle of G_{ijk} and $\boldsymbol{p}'$ is the position vector on L_2 after $\boldsymbol{p}$ is moved. The relationship between $\boldsymbol{p}$ and $\boldsymbol{p}'$ is shown in Fig. 5-1(i), where $\boldsymbol{n}$ is the normal vector of G_{ijk}. Thus, G_{ijk} is moved a distance d along the normal vector $\boldsymbol{n}$ to L_2, as shown in Fig. 5-1(j), where the new position is $\boldsymbol{g}'_{ijk}$. The grey circles with arrows in Fig. 5-1(k) are the dummy particles that have only one neighboring particle moved to L_1. The arrows are the direction of motion. The other two dummy particles without arrows shown in Fig. 5-1(k) have more than one neighboring particle moved to L_1.

If the dummy particle G_{ijk} has multiple neighboring particles located on L_1 and one of the distances between two neighboring particles on L_1 is less than L_0, G_{ijk} is not moved and is considered to be on L_2. If the dummy particle G_{ijk} has multiple neighboring particles and all the distances between all pairs of neighboring particles on L_1 are greater than or equal to L_0, G_{ijk} is moved as described by Eq. (5-4), where d is the shortest distance of the moved neighboring dummy particles. Fig. 5-1(l) shows a dummy particle G_{ijk} that has three neighboring particles G_{i-1jk}, G_{ij-1k}, and G_{ij+1k} that are moved to L_1. Because the distance between any two neighboring particles is larger than L_0, intial average distance, G_{ijk} is moved as on L_2. According to Eq. (5-5), the moving distance of G_{ijk} is zero because the moving distance of G_{i-1jk} is zero. Regarding another dummy particle G_{ijk}, as shown in Fig. 5-1(m), two neighboring particles G_{i-1jk} and G_{ij+1k} are moved to L_1. Because $|G_{i-1jk} - G_{ij+1k}| < L_0$, G_{ijk} is directly considered as being on L_2. Thus, L_2 is constructed as shown in Fig. 5-1(n), where the circles on the second line are the second layer.

The other layers can be constructed sequentially in the same manner as L_2. The third layer is constructed as shown in Figs. 5-1(o) and (p).

5.2 Adjustment of collision coefficients

In the polygon boundary condition proposed by Harada et al.[41], the collision model is used. The collision model mainly takes action when fluid particles are surface particles. Because the pressure is zero for surface particles, excessively close distances between particles do not cause the pressure gradient to change. Thus, the collision model adjusts the distances between these fluid particles. When the distances between the fluid particles are smaller than $\alpha_1 L_0$, the following repulsive velocity is enforced:

$$u' = -\alpha_2 u \qquad (5-6)$$

where u and u' are the velocity before and after the collision, respectively. To obtain the optimal collision model, Lee et al.[82] investigated this collision model. The coefficients α_1 and α_2 are set as 0.9 and 0.2, respectively. Here, we also use these collision coefficients to increase the spatial stability in simulations.

5.3 Simulation results

In this section, a boundary particle arrangement (BPA) technique coupled with the PBC is compared with the PBC without the BPA technique and the original MPS method[40] using two examples that include flat surfaces, right angles, slopes, and curved surfaces to demonstrate the performance of the proposed method with regard to defining the wall weight function in arbitrary geometries.

5.3.1 Dam break with a wedge

BPA is proposed to improve the wall weight function near non-planar polygons. Thus, we compare the particle number densities of the PBC with and without BPA near the slope in the water tank in Fig. 5-2. The initial setup of the investigated three-dimensional geometry is shown in Fig. 5-2. The length, width, and height of the water tank are 1.68, 0.72, and 1.4 m, respectively. The initial length and height of the fluid are 0.72 and 0.56 m, respectively. There is a wedge inside the water tank. The left plane of the wedge is a slope, and the right plane of the wedge is perpendicular to the bottom surface of the water tank. Table 5-1 shows the calculation parameters.

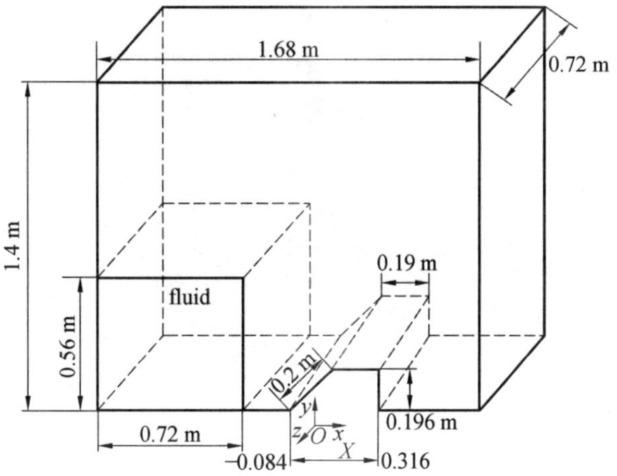

Fig. 5-2 Sketch of a water tank with a wedge

Table 5-1 Calculation parameters

Initial condition	Value	Unit
Diameter of fluid particle L_0	0.04	m
Influence radius r_e	3.1 L_0	m
Grid spacing using PBC with BPA	0.04	m
Grid spacing using PBC without BPA	0.04	m
Density ρ	1 000	kg/m^3
Number of fluid particles	4 284	

Fig. 5-3 shows the initial arrangement of the uniform grid near the wedge using BPA. The silver particles represent the vertices of polygons, and the black particles represent the initial grid points that are outside of or on polygons near the wedge. The boundary particles are generated along polygons by BPA, as depicted in Fig. 5-4, where the silver particles represent the vertices of polygons. Three layers of boundary particles are generated along the wedge. If the boundary particles are not properly located on the wall boundaries, the motion of the fluid particles will be dramatically disturbed. However, when using BPA, it is expected that the motion of the fluid particles is smooth.

Fig. 5-3 The initial arrangement of grid points near the wedge

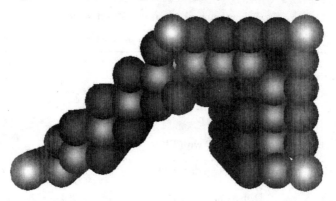

Fig. 5-4 The initial generated boundary particles near the wedge

Fig. 5-5 shows the distributions of fluid particles are at distances less than or equal to L_0 from the wedge at $t = 1.02$ s. The color of the particles represents the particle number density. The silver particles represent the vertices of polygons.

To clarify the differences in the particle number density with and without BPA, the distribution of the surface particles are compared at $t = 1.02$ s, as shown in Fig. 5-6. The black particles represent the free surface particles ($n_i < 0.97n^0$), and the grey ones are inner particles. Almost half of the fluid particles in the case without BPA are free surface particles, as shown in Fig. 5-6(a), which causes the motion of the fluid particles to be disturbed near the wedge. Conversely, in the case with BPA, most of the fluid particles near the wedge are not free surface particles, as shown in Fig. 5-6(b). Thus, the motion of the fluid particles is not disturbed near the wedge using the PBC with BPA.

Chapter 5 Boundary particle arrangement technique in polygon wall boundary condition

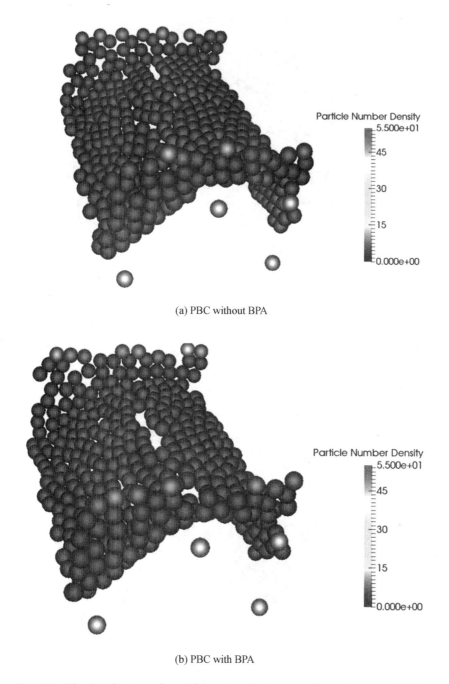

(a) PBC without BPA

(b) PBC with BPA

Fig. 5-5　The distributions of particle number density near the wedge at $t = 1.02$ s

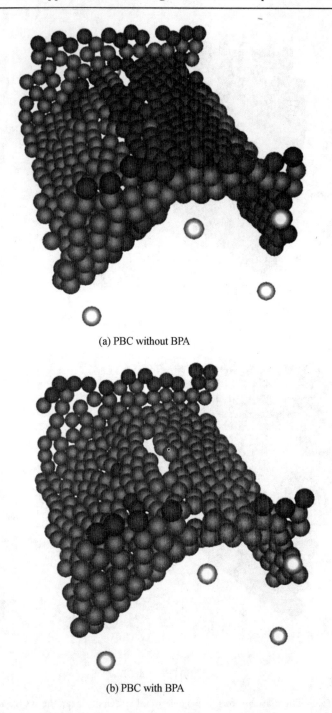

(a) PBC without BPA

(b) PBC with BPA

Fig. 5-6 The distributions comparison of surface particles near the wedge at $t = 1.02$ s

To quantitatively compare the difference between using the PBC with and without BPA, the distributions of the particle number densities of the fluid particles near the wedge for the two cases are compared, as shown in Fig. 5-7, where the horizontal axis represents the x-coordinate of the wedge, as defined in Fig. 5-2. Black lines represent n^0. grey points demonstrate the particle number density of the fluid particles. The grey points that lie far below n^0 in Fig. 5-7 are the fluid particles that are regarded as free surface particles, as shown in Fig. 5-6. The distribution of the particle number density using the PBC with BPA, shown in Fig. 5-7(b), is better than that of the PBC without BPA, shown in Fig. 5-7 (a). In Fig. 5-7 (a) there exist 432 fluid particles at distances less than or equal to L_0 from the wedge, whereas there are 378 such fluid particles in Fig. 5-7 (b). This illustrates that clustering of the fluid particles occurs near the wedge using the PBC with BPA. Thus, BPA can suppress the clustering of the fluid particles near non-planar polygons.

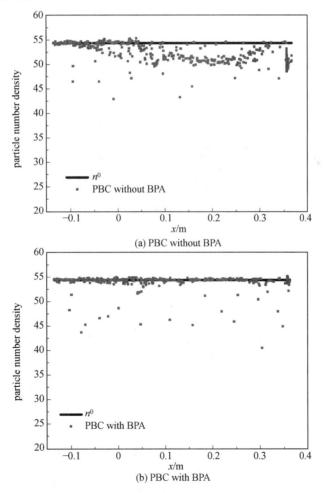

Fig. 5-7 The distributions of particle number densities near the wedge at $t = 1.02$ s

Snapshots of the simulation at $t = 1.02$ s are shown in Fig. 5-8. The color represents the particle number density. Although the snapshots of the two methods are similar, the PBC with BPA provides a better particle number density distribution near the non-planar polygons than does the PBC without BPA.

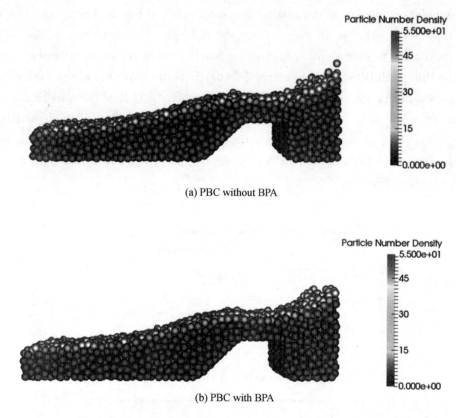

Fig. 5-8　Snapshots of a dam break with a wedge at $t = 1.02$ s

To confirm the effectiveness of BPA, the results of the PBC with and without BPA are compared at another time $t = 1.52$ s, as shown in Figs. 5-9~5-12. The simulation results at 1.52 s show the same tendencies as those at 1.02 s. The fluid particles at distances less than or equal to L_0 from the wedge are shown in Fig. 5-9, where the color represents the particle number density. Although the distributions of the particle number densities in Fig. 5-9 are intuitively similar, there are more free surface particles near the wedge using the PBC without BPA than with BPA, as shown in Fig. 5-10. There are 490 fluid particles near the wedge using the PBC without BPA, whereas there are 366 fluid particles using the PBC with BPA. Thus, fluid particles still cluster near the wedge because of the inaccurate particle number densities near the non-planar polygons. Fig. 5-11 compares the distributions of the particle number densities of the fluid particles using the PBC with and without BPA. The PBC with BPA provides a better distribution of the particle number density than does the PBC without BPA.

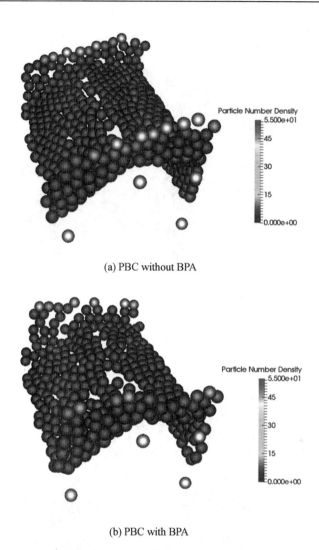

(a) PBC without BPA

(b) PBC with BPA

Fig. 5-9 Snapshots of dam break with a wedge at $t = 1.52$ s

Finally, the root mean square (RMS) respect to $n_i - n^0$ is compared at $t = 1.02$ and 1.52 s. The result is shown in Fig. 5-13. The RMS of the PBC with BPA is smaller than that without BPA at both time steps. This means that the particle number density is more accurately calculated near n^0 using the PBC with BPA.

(a) PBC without BPA

(b) PBC with BPA

Fig. 5-10　Comparison of surface particles near the wedge at $t = 1.52$ s

Chapter 5　Boundary particle arrangement technique in polygon wall boundary condition

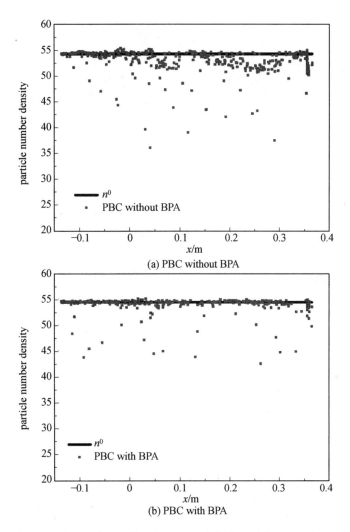

Fig. 5-11　Distributions of particle number densities of fluid particles near the wedge at $t = 1.52$ s

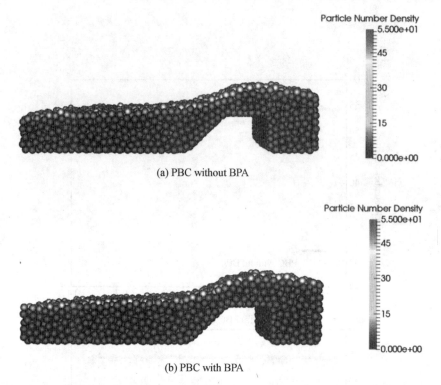

(a) PBC without BPA

(b) PBC with BPA

Fig. 5-12　Snapshots of dam break with a wedge at $t = 1.52$ s

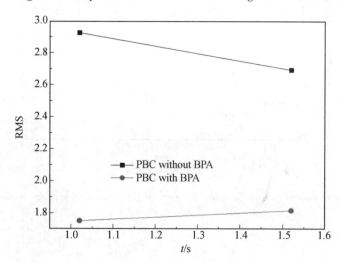

Fig. 5-13　Comparison of RMS with respect to $n_i - n^0$ at $t = 1.02$ and 1.52 s

5.3.2 Simulation of complex geometry

Next, the particle number density near curved boundaries is investigated to show the generality of BPA in improving the wall weight function. The investigated three-dimensional geometry is represented by polygons in Fig. 5-14. This geometry is composed of 481 vertices, which constitute 954 triangular polygons. Table 5-2 shows the scales of the geometry. Table 5-3 shows the calculation parameters. The boundary particles are located on the polygons using the PBC with BPA, as shown in Fig. 5-15. Three layers of boundary particles are arranged. The boundary particles are uniformly distributed on the polygons.

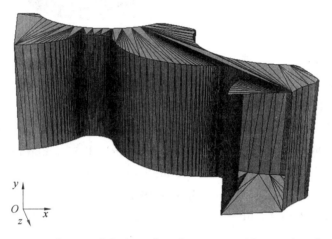

Fig. 5-14 Polygons of the investigated geometry with curved surfaces

Table 5-2 The scales of the geometry

Axis	Range	Unit
x	0.0~0.008 65	m
y	0.001~0.004 56	m
z	$-0.008\ 7$~0.0	m

Table 5-3 The calculation parameters

Initial condition	Value	Unit
Diameter of fluid particle L_0	0.000 1	m
Influence radius r_e	$3.1L_0$	m
Grid spacing using PBC with BPA	0.000 1	m
Grid spacing using PBC without BPA	0.000 1	m
Density ρ	1 000	kg/m^3
Number of fluid particles	80 586	

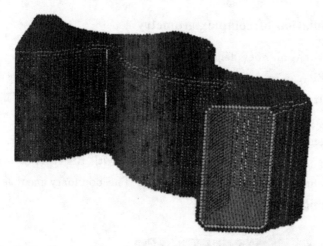

Fig. 5-15 Generated boundary particles along the curved boundaries

To simulate the complex geometry, the inlet and outlet boundary conditions are given at the left and right entries of the geometry, respectively, as shown in Fig. 5-15. Regarding the inflow and outflow boundaries, the periodic boundary condition[2] and standard inlet and outlet boundary conditions[101] are commonly used. However, these methods can only simulate the situation in which the inflowing and outflowing particle numbers are equal. Federico et al.[57] simulated a free-surface channel flow with improved inlet and outlet boundary conditions. Shakibaeinia et al.[50] proposed a general inlet and outlet boundary condition with a recycling strategy. For the complex geometry shown in Fig. 5-16, the scale of the inflow boundary is different from that of the outflow boundary. Thus, the recycling strategy proposed by Shakibaeinia et al.[50] is adopted to satisfy the mass conservation law. The inflow velocity is designated as $u_x = 1, u_y = 0, u_z = 0$ m/s. The outflow velocity must be carefully controlled to ensure the flow does not have a free surface. According to our study, the outflow velocity is not only affected by the inflow velocity but also by the shape of the geometry. If the geometry is not regular, the outflow velocity is affected by changes in the shape nearest to the outflow boundary, which is the narrow neck close to the outflow boundary, namely the interval $z \in [-0.003, -0.002]$. To determine the outflow velocity, the weighted average velocity in this interval is calculated by

$$\boldsymbol{u}'_i = \frac{\sum\limits_{j \in \text{fluid}} u_j n_j}{\sum\limits_{j \in \text{fluid}} n_j} \tag{5-7}$$

where $\boldsymbol{u}'_i$ is the estimated velocity of the ith fluid particle near the outflow boundary; u_j is the velocity of the jth fluid particle along z direction belonging to the interval $z \in [-0.003, -0.002]$, and n_j is the particle number density of the jth fluid particle. Depending on the geometry, the outflow velocity of the ith fluid particle is adjusted by

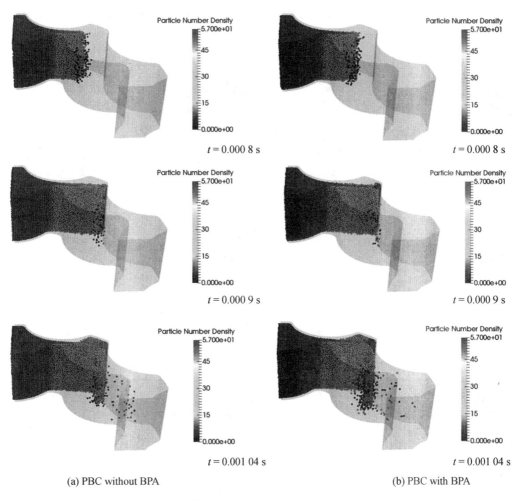

(a) PBC without BPA (b) PBC with BPA

Fig. 5-16 Side view of complex geometry at $t = 0.000\,8$, $0.000\,9$, and $0.001\,04$ s

$$\boldsymbol{u}_i = \alpha \boldsymbol{u}'_i \tag{5-8}$$

where α is an adjustment coefficient. In this research, α is set as 0.9. Snapshots are shown in Fig. 5-16 at three times $t = 0.000\,8$, $0.000\,9$, and $0.001\,04$ s. The top-down views of these snapshots are shown in Fig. 5-17. The color of the particles represents the particle number density. More yellow particles can be seen near the curved boundaries when the PBC is used without BPA than with BPA. The PBC with BPA provides a better distribution of the particle number density than does the PBC without BPA. The particle number density of the fluid particles at distances less than or equal to L_0 from a polygon at $t = 0.000\,8$, $0.000\,9$, and $0.001\,04$ s are compared in Fig. 5-18. The grey points represent the particle number density of the fluid particles, and the black lines are n^0. At $t = 0.000\,8$ s the particle number density in both methods from $x = 0.001\,7$ to $0.002\,0$ m shows a large deviation from n^0 because these fluid particles are at the front and thus are regarded as being on the free surface. With the exception of this phenomenon, the distributions of the particle number densities using the PBC with BPA are better than those of the PBC without BPA.

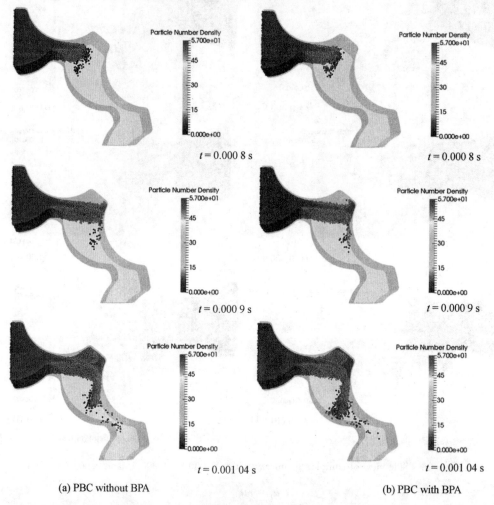

(a) PBC without BPA   (b) PBC with BPA

Fig. 5-17 Top-down view of complex geometry at $t = 0.000\,8$, $0.000\,9$, and $0.001\,04$ s

The RMS of $n_i - n_0$ is compared at three time steps for the PBC with and without BPA, as shown in Fig. 5-19. The RMS for the PBC with BPA is smaller than that without BPA. The comparison of the particle number density near the curved surface demonstrates that BPA can dramatically improve the wall weight function near curved boundaries.

The particle number densities of fluid particles near the bottom boundary using the PBC with and without BPA are quantitatively compared at coordinates in the range $y \in [0.001, 0.001\,5]$, i.e., particles within a distance of $5L_0$ from the bottom surface are considered. The simulation results of the inlet and outlet boundary conditions by the PBC with and without BPA at $t = 5$ s are shown in Fig. 5-20. The color represents the particle number density. The particle number density distributions are shown in

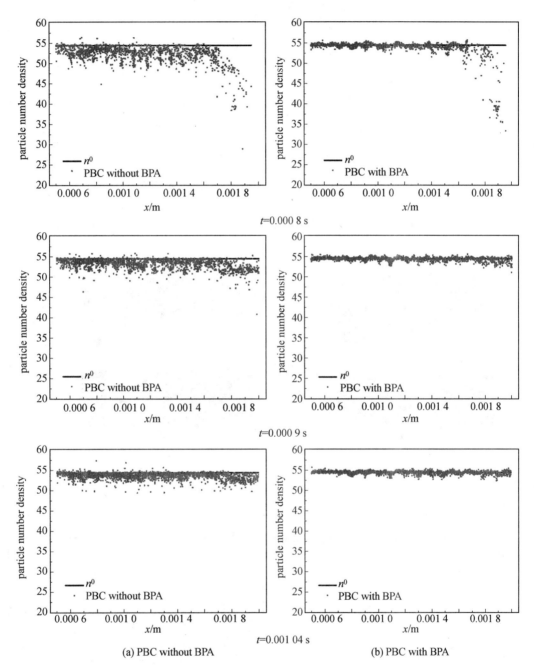

Fig. 5-18 Distributions of particle number densities of complex geometry near curved surface at $t = 0.0008$, 0.0009, and 0.00104 s

Fig. 5-21. The PBC with BPA again presents a better particle number density distribution than does the PBC without BPA.

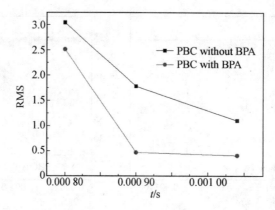

Fig. 5-19 Comparison of RMS with respect to $n_i - n^0$ at $t = 0.000\ 8,\ 0.000\ 9,\ 0.001\ 04$ s

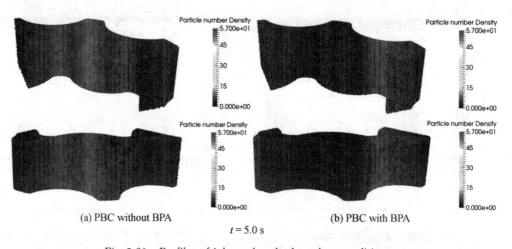

(a) PBC without BPA (b) PBC with BPA

$t = 5.0$ s

Fig. 5-20 Profiles of inlet and outlet boundary conditions

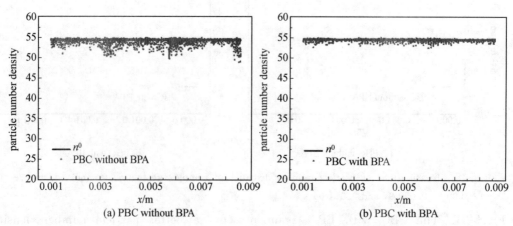

(a) PBC without BPA (b) PBC with BPA

Fig. 5-21 Comparison of particle number density distributions

5.4 Summary

In this section, we developed an initial boundary particle arrangement technique coupled with the PBC to improve the wall weight function near slopes and curved surfaces. Two uniform grids are utilized in the proposed technique. The boundary particles are generated at the grid points in the first uniform grid and moved to the boundary. The second uniform grid stores the information of the PBC. The generated boundary particles are used in the initial procedure, and the stored information is used in the main loop. Thus, the efficiency of the PBC is retained, and three-dimensional complex geometries can be treated using the proposed technique.

To show the general applicability of the BPA technique to complex geometries, the particle number density distributions near slopes and curved surfaces are compared for cases with and without BPA. The simulation results demonstrate that the BPA technique can effectively improve the particle number density distribution near non-planar polygons. Although the particle number density has been improved with the BPA technique, the pressure oscillation is still large in the polygon boundary condition. Next, we plan to improve the pressure stability with the BPA technique and apply it to further complex geometries.

Chapter 6   Improvement of pressure distribution in polygon wall boundary condition

6.1   Research progress

In Chapter 5 the particle number density has been improved by coupling the improved wall weight function introduced in Chapter 4 with the initial boundary particle arrangement (BPA) technique introduced in Chapter 5. However, the pressure oscillation in the polygon wall boundary condition is still severe, which causes unreal simulation results.

The improvement of pressure distribution is not only important in polygon wall boundary condition, but also a common issue in particle methods. In MPS method[40], Ikeda[102] tried small time step to reduce the pressure oscillation. Hibi[103] introduced two post-processing to suppress the pressure oscillation. Ataie and Farhadi[49] validated different kernel functions to improve the stabilization of MPS method. However, these methods cannot solve the pressure oscillation fundamentally. To stabilize the pressure distribution, Tanaka and Masunaga[104] proposed a new source term combining divergence of velocity with variation of particle number density to suppress the pressure oscillation and improve the efficiency. This method can effectively suppress the pressure oscillation. Koh et al.[105] also employed this source term in consistent particle method (CPM)[106] to improve the simulation of floating body. Khayyer and Gotoh[107] proposed a symmetric gradient discretization formulation to conserve momentum of MPS method. Kondo and Koshizuka[108] derived a general form of source term to suppress pressure oscillation. In essence methods used in [104] and [108] are the same algorithms. Khayyer and Gotoh[109] proposed an improved Poisson pressure equation coupled with the momentum conservative gradient model[107] to predict wave impact pressure. Lee et al.[82] improved the spatial stability through revised collision coefficients.

In SPH[38, 39] method, Colagrossi and Landrini[65] reduced the numerical noise with MLS integral interpolation based on the momentum preservation SPH[60]. Molteni and Colagrossi[110] proposed a density diffusion term to improve the computation of pressure distribution. Hosseini and Feng[111] improved the accuracy of simulations by enforcing accurate nonhomogeneous Neumann boundary condition in incompressible SPH (ISPH). A particle shifting method[112, 113] was proposed to improve the numerical stability and prevent anisotropic distributions of particles. Adami et al.[66] proposed a simple formulation based on local force balance to treat complex geometry and improve pressure distributions.

To obtain smooth distribution of pressure, boundary condition is very important. In all these improvements introduced above, dummy particles are utilized i. e. wall boundaries are discretized as layers of boundary particles to enforce repulsive force to the fluid particles. Although stabilization of pressure can be improved with ghost particles, layers of boundary particles are difficult to be arranged with strictly equal distance especially in complex geometries as we introduced in Section 3.1.

In polygon wall boundary condition, Yamada et al.[97] proposed an explicit polygon boundary condition based on the MPS method with the same idea as the method by Harada et al.[41]. This method still suffers from the strong pressure oscillation near the corners and angles. To suppress the unphysical motions of fluid particles near the polygons, Mitsume et al.[98] developed a new explicit polygon boundary condition which calculates the closest distance with a search algorithm without using the distance function. Mirror particle and a boundary force are employed as the mixed boundary condition. Although this method is more accurate than the method by Yamada et al.[97], the inaccurate boundary force and time consuming distance calculation especially in complex geometries not only cause strong pressure oscillation, but also reduce the efficiency.

In this chapter we clarify the problems of the Poisson's equation in the polygon boundary condition, and then introduce the source term proposed by Tanaka and Masunaga[104] to suppress the pressure oscillation of fluid particles far from the polygons. To supplement the wall weight function, the BPA technique is utilized. A proportion factor of the particle number density is introduced in the source term to accurately represent the contribution of the fluid and wall parts of the Poisson's equation when fluid particles are close to the polygons. To smooth the effect caused by the non-uniform generated boundary particles, the weighted average of the particle number density is introduced in the source term to suppress the numerical oscillation of pressure. The asymmetric gradient model derived by Khayyer and Gotoh[107] is adopted to further improve the pressure distribution. The proposed method maintains the simplicity of the polygon boundary condition and the pressure oscillation caused by the polygons can be effectively suppressed. The complex geometries can also be simulated with high efficiency.

Four 3-D examples are tested to illustrate the performance of the proposed method. The hydrostatic simulation is tested to illustrate the improvement of the pressure calculations. The dam break simulation is compared between the proposed method and other models to verify the effectiveness of the proposed method to suppress the pressure oscillation near polygons. Finally, two complex geometries are simulated to demonstrate the general applicability of the proposed method to improve the pressure distribution in complex geometries.

6.2 Improved particle-polygonal meshes interaction models

In this section the problems of the Poisson's equation in the polygon boundary condition is analyzed. The new fluid and wall parts of the Poisson's equation are proposed coupled with the BPA technique to suppress the pressure oscillation in polygon boundary condition. The asymmetric gradient model proposed by Khayyer and Gotoh[107] is also adopted to further improve the pressure calculation and suppress the pressure oscillation near the polygons.

6.2.1 Re-derivation of polygon wall boundary condition

Although in Section 3.3.2 the discretization equations used in the polygon boundary condition have been introduced, the necessary discretization equations are summarized here for easily illustrating the contents of this chapter.

The discretization of the gradient in polygon boundary condition is formulated as

$$\langle \nabla P \rangle_i = \langle \nabla P \rangle_{if} + \langle \nabla P \rangle_{iw} \tag{6-1}$$

$$\langle \nabla P \rangle_{if} = \frac{d}{n_0} \sum_{j \neq i} \left[\frac{P_j - \hat{P}_i}{|r_j - r_i|^2} (r_j - r_i) w(|r_j - r_i|) \right] \tag{6-2}$$

$$\langle \nabla P \rangle_{iw} = \frac{\rho |dr|}{\Delta t^2} \frac{r_{iw}}{|r_{iw}|} \tag{6-3}$$

where $\langle \nabla P \rangle_{if}$ and $\langle \nabla P \rangle_{iw}$ are the fluid and wall parts of pressure gradient of fluid particle i; $\hat{P}_i$ is the minimum pressure in the effective radius r_e. The fluid part is the same as the original MPS, and the wall part is transformed to the equation of distance vector. dr and r_{iw} are the distance vector from fluid particle to the average position and to the polygon wall boundary, respectively as introduced in Section 3.3.1. When the distance of fluid particle i to the wall boundaries are less than L_0, Eq. (6-3) takes effect.

The Poisson's equation used in the polygon boundary condition is also divided into two parts and evaluated by

$$\langle \nabla^2 P \rangle_i = \langle \nabla^2 P \rangle_{if} + \langle \nabla^2 P \rangle_{iw} = -\frac{\rho}{\Delta t^2} \frac{n_i^* - n^0}{n^0} \tag{6-4}$$

$$\langle \nabla^2 P \rangle_{if} = \frac{2d}{\lambda n^0} \sum_{j \neq i, j \in \text{fluid}} \left[(P_j - P_i) w(|r_j - r_i|) \right] \tag{6-5}$$

$$\langle \nabla^2 P \rangle_{iw} = \frac{2\rho |dr| |r_{iw}|}{\lambda \Delta t^2} \tag{6-6}$$

where $\langle \nabla^2 P \rangle_{if}$ is the fluid part of Poisson's equation and $\langle \nabla^2 P \rangle_{iw}$ is the wall part; n_i^* is the intermediate particle number density; $\langle \nabla^2 P \rangle_{iw}$ is calculated when the fluid particles have the distance to the polygons less than L_0; λ is the Laplacian model coefficientthe same as MPS method[40].

The Dirichlet boundary condition is used to detect surface particles. The pressure of particles is set to zero when the particle number density of the fluid particle satisfies the equation

$$n_i < \beta n^0 \tag{6-7}$$

where β is a parameter for surface detection. In this study $\beta = 0.97$ is adopted.

The same as Chapter 5, the coefficients α_1 and α_2 used in the collision model are also assigned as 0.9 and 0.2, respectively.

6.2.2 Problem of present source term in the polygon wall boundary condition

In the polygon boundary condition, the Laplacian namely the left side of the source term in Eq. (6-4) is divided into the fluid and wall parts. Since the wall part only relates to the distance which is determined before calculating the Poisson's equation, the Poisson's equation is calculated by

$$\langle \nabla^2 P \rangle_{if} = -\frac{\rho}{\Delta t^2} \frac{n_i^* - n^0}{n^0} - \frac{2\rho |dr| |r_{iw}|}{\lambda \Delta t^2} \tag{6-8}$$

The source term in Eq. (6-8) employs the variation of the particle number density to the standard particle number density. Using this source term, the error is not accumulated with the time. However, pressure oscillation is drastic in terms of space and time as explained in[104]. Thus, this source term itself causes the numerical oscillation of pressure.

When the fluid particle has the distance to the polygons less than L_0 the repulsive force is enforced from the polygons. Thus, the wall part of Poisson's equation should become large when the fluid particle approaches the polygons i.e., Eq. (6-6) should become large when $|r_{iw}|$ decreases. However, Eq. (6-6) does not increase all the time as $|r_{iw}|$ decreases as shown in Fig. 6-1 where L_0 is chosen as 0.04 m. In fact, the wall part of the Poisson's equation first increases and then decreases as $|r_{iw}|$ decreases. Thus, inconsistent contribution of the polygons also causes the pressure oscillation.

Finally, the repulsive force from the polygons to the fluid particle is imposed when $|r_{iw}| < L_0$ in Eq. (6-6). The contribution from the polygons to the fluid particles is neglected when $L_0 \leqslant |r_{iw}| < r_e$, which not only causes the pressure oscillation, but also affects the accuracy of the simulations. Thus, these three problems in the Poisson's equation lead to the drastic fluctuation of pressure.

6.2.3 Improvement of source term in the polygon wall boundary condition

Since the accuracy of the Poisson's equation directly effects the calculation of the pressure field, accurately calculating the Poisson's equation is very important. The

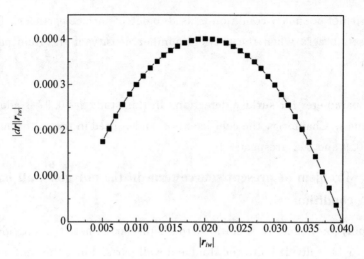

Fig. 6-1 Relationship between $|r_{iw}|$ and $|dr||r_{iw}|$

pressure oscillation is caused by the interaction between fluid particles without the influence by polygons and the interaction between fluid particle and polygons. To suppress the pressure oscillation of the fluid particles far from the polygons, several methods can be adopted such as the improved source term proposed by Tanaka and Masunaga[104] or by Kondo and Koshizuka[108]. In this research, we choose improved source term proposed by Tanaka and Masunaga[104] because only one parameter need to be adjusted. The source term that combines the divergence of velocity and the variation of the particle number density proposed by Tanaka and Masunaga[104] is introduced

$$\nabla^2 P_i = (1-\gamma)\frac{\rho}{\Delta t}\nabla \cdot u_i^* + \gamma \frac{\rho}{\Delta t^2}\frac{n^0 - n^k}{n^0} \tag{6-9}$$

where the parameter γ is in the range of $0.01 < \gamma < 0.05$ to keep the volume conservation. Thus, the Poisson's equation in polygon boundary condition can be replaced by

$$\langle \nabla^2 P \rangle_{if} = (1-\gamma)\frac{\rho}{\Delta t}\nabla \cdot u_i^* + \gamma \frac{\rho}{\Delta t^2}\frac{n^0 - n^k}{n^0} - \frac{2\rho |dr||r_{iw}|}{\lambda \Delta t^2} \tag{6-10}$$

where the divergence of velocity at particle i is discretized as

$$\langle \nabla \cdot u \rangle_i = \frac{d}{n^0}\sum_{j \neq i}\frac{(u_j - u_i)\cdot(r_j - r_i)}{|r_j - r_i|^2}w(|r_j - r_i|) \tag{6-11}$$

Divide the divergence of velocity into the fluid and wall parts

$$\langle \nabla \cdot u \rangle_i = \langle \nabla \cdot u \rangle_{if} + \langle \nabla \cdot u \rangle_{iw} \tag{6-12}$$

where $\langle \nabla \cdot u \rangle_{if}$ and $\langle \nabla \cdot u \rangle_{iw}$ are the fluid and wall parts of fluid particle i, respectively. The wall part can be formulated as

$$\langle \nabla \cdot u \rangle_{iw} = \frac{d}{n^0}\frac{u_{iw}\cdot r_{iw}}{|r_{iw}|^2}Z_i \tag{6-13}$$

where $u_{iw} = u_w - u_i$ and r_{iw} can be interpolated from the normal vectors and closest distances stored at the grid points in the second uniform grid. Unfortunately, due to the interpolation error, r_{iw} becomes very large when particle i approaches the polygons, which leads to very large $\langle \nabla \cdot \boldsymbol{u} \rangle_{iw}$ and finally the simulation diverges. To prevent numerical divergence, the source term composed by the variations of the particle number densities is employed because the particle number density can be calculated accurately by BPA technique. Thus, the new source term is formulated as

$$\nabla^2 P_i = (1-\gamma) \frac{\rho}{\Delta t^2} \frac{n_i^k - n_i^*}{n^0} + \gamma \frac{\rho}{\Delta t^2} \frac{n^0 - n_i^k}{n^0} \qquad (6\text{-}14)$$

The Laplacian can be divided into the fluid and wall parts. Substitute Eq. (6-3) into the wall part. The fluid part can be calculated by

$$\langle \nabla^2 P \rangle_{if} = (1-\gamma) \frac{\rho}{\Delta t^2} \frac{n_i^k - n_i^*}{n^0} + \gamma \frac{\rho}{\Delta t^2} \frac{n^0 - n_i^k}{n^0} - \frac{2\rho |dr| |r_{iw}|}{\lambda \Delta t^2} \qquad (6\text{-}15)$$

When fluid particle i has the distance to the polygons less than L_0, Eq. (6-15) is calculated. Otherwise, Eq. (6-14) is utilized to compute the pressure of fluid particle i. Using Eq. (6-15) can improve the pressure calculations near the polygons. However, the pressure oscillation becomes more drastic than using Eq. (6-8). The reason is that the wall part of the Poisson's equation is not accurate. Although the source term is changed, the wall part of the Poisson's equation does not change with the source term, which causes stronger pressure oscillation.

To accurately calculate the contribution of the fluid and wall parts of the Poisson's equation, we first discretize the Laplacian in Eq. (6-14) into two parts

$$\langle \nabla^2 P \rangle_{if} + \langle \nabla^2 P \rangle_{iw} = (1-\gamma) \frac{\rho}{\Delta t^2} \frac{n_i^k - n_i^*}{n^0} + \gamma \frac{\rho}{\Delta t^2} \frac{n^0 - n_i^k}{n^0} \qquad (6\text{-}16)$$

As we seen, since the source term in Eq. (6-16) is composed by the variation of the particle number density, we can assume that the contribution of the fluid and wall parts to the source term is reflected by the ratio of each part in the particle number density. Thus, we divide the source term into the fluid and wall parts by the contribution of each part to particle number density. A proportion factor of the particle number density is introduced into the source term and the Poisson's equation can be represented by

$$\langle \nabla^2 P \rangle_{if} + \langle \nabla^2 P \rangle_{iw}$$
$$= \frac{n_{if}^*}{n_i^*} \{ (1-\gamma) \frac{\rho}{\Delta t^2} \frac{n_i^k - n_i^*}{n^0} + \gamma \frac{\rho}{\Delta t^2} \frac{n^0 - n_i^k}{n^0} \} + \frac{n_{iw}^*}{n_i^*} \{ (1-\gamma) \frac{\rho}{\Delta t^2} \frac{n_i^k - n_i^*}{n^0} + \gamma \frac{\rho}{\Delta t^2} \frac{n^0 - n_i^k}{n^0} \}$$
$$(6\text{-}17)$$

where n_{iw}^* is the wall weight function of fluid particle i after calculating the explicit process and $n_{if}^* = n_i^* - n_{iw}^*$. Since the source term of Eq. (6-17) can be regarded as the respective contribution of the fluid and wall parts, the Laplacian of fluid and wall parts can be directly represented by

$$\langle \nabla^2 P \rangle_{if} = \frac{n_{if}^*}{n_i^*} \{ (1-\gamma) \frac{\rho}{\Delta t^2} \frac{n_i^k - n_i^*}{n^0} + \gamma \frac{\rho}{\Delta t^2} \frac{n^0 - n_i^k}{n^0} \} \qquad (6\text{-}18)$$

$$\langle \nabla^2 P \rangle_{iw} = \frac{n_{iw}^*}{n_i^*} \{ (1-\gamma) \frac{\rho}{\Delta t^2} \frac{n_i^k - n_i^*}{n^0} + \gamma \frac{\rho}{\Delta t^2} \frac{n^0 - n_i^k}{n^0} \} \qquad (6\text{-}19)$$

Using the equations above, the fluid and wall parts of the Poisson's equation are determined by the contribution of the particle number density. The wall part of the Poisson's equation can changes with the source term rather than a simple distance function independent of the source term.

To avoid slight fluctuation of the particle number density caused by the uneven boundary particles generated by BPA technique, the weighted average of the particle number density is used to replace the second n_i^k on the right side of Eqs. (6-18) and (6-19). Thus, the fluid and wall parts of the Poisson's equation are formulated respectively as

$$\langle \nabla^2 P \rangle_{if} = \frac{n_{if}^*}{n_i^*} \{ (1-\gamma) \frac{\rho}{\Delta t^2} \frac{n_i^k - n_i^*}{n^0} + \gamma \frac{\rho}{\Delta t^2} \frac{n^0 - ns_i^k}{n^0} \} \qquad (6\text{-}20)$$

$$\langle \nabla^2 P \rangle_{iw} = \frac{n_{iw}^*}{n_i^*} \{ (1-\gamma) \frac{\rho}{\Delta t^2} \frac{n_i^k - n_i^*}{n^0} + \gamma \frac{\rho}{\Delta t^2} \frac{n^0 - ns_i^k}{n^0} \} \qquad (6\text{-}21)$$

where ns_i^k is calculated by

$$ns_i^k = \frac{\sum_{j \neq i, j \in \text{fluid}} n_j^k w(r_{ij})}{\sum_{j \neq i, j \in \text{fluid}} w(r_{ij})} \qquad (6\text{-}22)$$

when $r_{iw} < r_e$, Eq. (6-22) is employed to calculate the pressure of fluid particles. When $r_{iw} \geqslant r_e$, the influence caused by the wall boundary disappears. The Poisson's equation is turned into

$$\langle \nabla^2 P \rangle_{if} + 0 = (1-\gamma) \frac{\rho}{\Delta t^2} \frac{n_i^k - n_i^*}{n^0} + \gamma \frac{\rho}{\Delta t^2} \frac{n^0 - ns_i^k}{n^0} \qquad (6\text{-}23)$$

where $\langle \nabla^2 P \rangle_{iw} = 0$.

6.2.4 Improvement of gradient model

In the polygon boundary condition, the pressure gradient model of the fluid and wall parts is calculated by Eqs. (6-2) and (6-3), respectively. However, Eqs. (6-2) and (6-3) do not conserve the momentum. Li et al.[72] use the linear momentum conservative gradient model proposed by Khayyer and Gotoh[107] to suppress the pressure oscillation in polygon boundary condition. In present study this pressure gradient model is also employed

$$\langle \nabla P \rangle_i = \frac{d}{n_0} \sum_{j \neq i} \left[\frac{(P_i + P_j) - (\hat{P}_i + \hat{P}_j)}{|r_j - r_i|^2} (r_j - r_i) w(|r_j - r_i|) \right] \qquad (6\text{-}24)$$

This pressure gradient model can obtain smooth pressure distribution. However, the pressure value of fluid particles is far larger than the real value. To solve this problem, the strictly derived gradient model is adopted

$$\langle \nabla P \rangle_i = \frac{d}{n_0} \sum_{j \neq i} \left[\frac{P_i + P_j - 2\hat{P}_i}{|\bm{r}_j - \bm{r}_i|^2} (\bm{r}_j - \bm{r}_i) w(|\bm{r}_j - \bm{r}_i|) \right] \quad (6\text{-}25)$$

Although this gradient model is not conservative, it can improve the pressure calculation near the polygons and suppress the pressure oscillation.

The polygon boundary condition can be derived using the new gradient model from the beginning. In the original MPS, the Navier-Stokes equation is divided into the explicit and implicit processes. The implicit process is represented by

$$\bm{u}' = -\frac{\Delta t}{\rho} \nabla P^{k+1} \quad (6\text{-}26)$$

Replace ∇P^{k+1} with discretization Eq. (6-25) and multiply Δt on both sides of Eq. (6-26). The distance vector equation can be derived

$$d\bm{r} = -\frac{d \Delta t^2}{\rho n_0} \frac{P_i + P_w - 2\hat{P}_i}{|\bm{r}_{iw}|^2} Z_i \bm{r}_{iw} \quad (6\text{-}27)$$

Vector $d\bm{r}$ and $\bm{r}_{iw}$ satisfy $\frac{d\bm{r}}{|d\bm{r}|} = -\frac{\bm{r}_{iw}}{|\bm{r}_{iw}|}$. Thus, the pressure can be obtained by

$$P_i + P_w - 2\hat{P}_i = \frac{\rho n_0 |d\bm{r}| |\bm{r}_{iw}|}{d \Delta t^2 Z_i} \quad (6\text{-}28)$$

Substitute Eq. (6-28) into Eq. (6-25). The new gradient model of wall part is obtained and it is the same as Eq. (6-6). In fact to any gradient model, the source term of the wall part is the same. Thus, we continue using Eq. (6-6) as the pressure gradient model of the wall part. The advantage of Eq. (6-6) is that it can transform the boundary condition to the distance function, which can be easily applied to the complex geometries.

6.2.5 Surface detection

In the polygon boundary condition, the surface particles are detected by Eq. (6-7) the same as original MPS. Tanaka and Masunaga[104] propose a mixed surface detection method that combines Eq. (6-7) with the number of neighbor particles to improve the surface detection

$$N_i < \beta' N_0 \quad (6\text{-}29)$$

where N_0 is the maximum number of neighboring particles within effective radius for fully submerged particles in the initial distribution; β' is the experience coefficient which is assigned as 0.8 in the present study; N_i is the number of neighboring particles and it is formulated as

$$N_i = \sum_{j \neq i} w^s(r_{ij}) \quad (6\text{-}30)$$

where w^s is formulated as

$$w^s(r_{ij}) = \begin{cases} 1 & (r_{ij} \leqslant r_e) \\ 0 & (r_{ij} > r_e) \end{cases} \quad (6\text{-}31)$$

Divide Eq. (6-30) into the fluid and wall parts

$$N_i = \sum_{j \neq i, j \in \text{fluid}} w^s(r_{ij}) + \sum_{k \neq i, k \in \text{boundary}} w^s(r_{ik}) \tag{6-32}$$

The number of the boundary particles can be computed by the generated boundary particles. In this research the mixed surface detection is also adopted to judge free surface. To improve the efficiency, it is the same as the computation of the wall weight function. If $r_{iw} \geqslant r_e$, boundary particles do not need to be calculated. Otherwise, boundary particles are calculated by the generated boundary particles.

6.3 Results and discussions

In this section four examples are presented to validate the performance of the proposed method to suppress the numerical oscillations of pressure. The first example shows the development of the hydrostatic pressure. The second example is the dam break simulation where the pressure profiles are compared by different source terms and gradient models. To test the robust of the proposed method, dam break with a wedge and a complex geometry are simulated to show the generality of the proposed method to improve the pressure distribution in complex geometries.

6.3.1 Hydrostatic pressure

To illustrate the accuracy of the pressure calculation, the hydrostatic pressure is simulated. The initial set-up is shown in Fig. 6-2. The length, width and height of the water tank are 1.28, 0.72, and 0.72 m, respectively. The height of the fluids is 0.52 m. The diameter of fluid particle $L_0 = 0.04$ m and the effective radius r_e is $3.1L_0$. The total number of fluid particles used for the simulation is 5 851. In all the simulations below, the gravity acceleration is 9.81 m/s^2, the collision coefficients employ 0.9 and 0.2, respectively, and all the simulations use Eq. (6-32) to judge the free surface.

Table 6-1 shows the various source terms and gradient models used in the hydrostatic simulation. Table 6-2 shows the correspondent equations used in table 6-1. The hydrostatic pressures of different cases are compared with the theory values at $t = 2.0$ s as shown in Fig. 6-3. To show the difference clearly, all cases are divided into two groups. Fig. 6-3 (a) compares the hydrostatic pressure between Case 01 and Case 03. Case 01 uses the Poisson's equation and the pressure gradient model proposed by Harada et al.[41]. Case 03 employs the proposed Poisson's equation and the linear momentum conservative gradient model proposed by Khayyer and Gotoh[107]. The hydrostatic pressures with Case 03 are far larger than that with Case 01 and the theory values because Eq. (6-24) is modified manually, which leads to the abnormal increase of the

Chapter 6 Improvement of pressure distribution in polygon wall boundary condition

particle number density and pressure of fluid particles. Thus, the linear momentum conservative gradient model cannot be used in the polygon boundary condition.

Table 6-1 Description of Poisson's equation and gradient models used in hydrostatic and dam break simulation

	Wall part of Poisson's equation	Fluid part of Poisson's equation	Fluid part of Gradient model
Case 01	Wall part by Harada et al.[41]	Source term of Koshizuka & Oka[40]- wall part by Harada et al.[41]	Fluid part by Harada et al.[41]
Case 02	Wall part by Harada et al.[41]	Source term of Tanaka & Masunaga[104]- wall part by Harada et al.[41]	Fluid part by Harada et al.[41]
Case 03	Wall part of proposed method	Fluid part of proposed method	symmetrical Gradient term by Khayyer & Gotoh[107]
Case 04	Wall part of proposed method	Fluid part of proposed method	Fluid part by Harada et al.[41]
Case 05	Wall part of proposed method without weighted average	Fluid part of proposed method without weighted average	asymmetric Gradient term by Khayyer & Gotoh[107]
Proposed method	Wall part of proposed method	Fluid part of proposed method	asymmetric Gradient term by Khayyer & Gotoh[107]

Table 6-2 Equations of Poisson's equation and gradient models used in hydrostatic and dam break simulation

	Wall part of Poisson's equation	Fluid part of Poisson's equation	Fluid part of Gradient model
Case 01	Eq. (6-6)	Eq. (6-5)	Eq. (6-2)
Case 02	Eq. (6-6)	Eq. (6-10)	Eq. (6-2)
Case 03	Eq. (6-21)	Eq. (6-20)	Eq. (6-24)
Case 04	Eq. (6-21)	Eq. (6-20)	Eq. (6-2)
Case 05	Eq. (6-19)	Eq. (6-18)	Eq. (6-25)
Proposed method	Eq. (6-21)	Eq. (6-20)	Eq. (6-25)

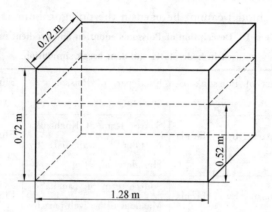

Fig. 6-2 Simulation condition of hydrostatic problem

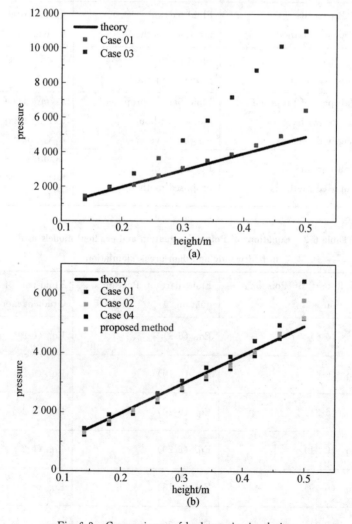

Fig. 6-3 Comparisons of hydrostatic simulations

Next the proposed method is compared with Case 01, Case 02, Case 04 and the theory values as shown in Fig. 6-3 (b). Case 02 uses the source term proposed by Tanaka and Masunaga[104]. Case 04 uses the proposed Poisson's equation but the pressure gradient model still employs the one used in the polygon boundary condition[41].

The pressure values in Case 01 becomes larger than the theory values with the increase of the depth. Case 02 has a better agreement with the theory values than Case 01 because of using the source term by Tanaka and Masunaga[104]. However, the pressure is still far larger than the theory values near the bottom. The pressure values with Case 04 have a larger difference from the theory values than Case 02 at most heights. However, near polygons Case 04 obtains better pressure than Case 02, which means the proposed Poisson's equation can obtain better pressure distribution near polygons. Using gradient model Eq. (6-25) the proposed method has a better agreement with the theory values than other Cases.

Fig. 6-4 shows the snapshots of the hydrostatic simulation between Case 01 and the proposed method at $t = 2.0$ s. The pressure distribution of Case 01 is not stable and the pressure oscillation is obvious. The proposed method has a better pressure distribution.

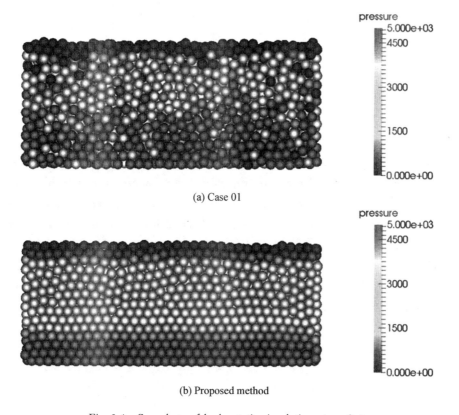

(a) Case 01

(b) Proposed method

Fig. 6-4 Snapshots of hydrostatic simulation at $t = 2.0$ s

6.3.2 Dam break simulation

In this section the classic dam break simulation is tested to show the improvement of the proposed method to reduce the pressure oscillation. The simulation condition is illustrated in Fig. 6-5. The length, width and the height of the water tank are 1.28 m, 0.72 m and 0.72 m, respectively. The length and height of the fluids are 0.56 m and 0.52 m, respectively. The total number of fluid particles is 3 094. The diameter of the fluid particle is 0.04 m and the influence radius $r_e = 3.1L_0$.

To validate the ability of the proposed method to suppress the pressure oscillation near polygons, three points A, B and C are tested on the polygons respectively. Point A is on the center of the bottom edge of the right surface. Point B is vertical to the edge where point A is located. The distance between point A and B is 0.04 m. Point C is on the intersection point of the back, right and bottom surfaces.

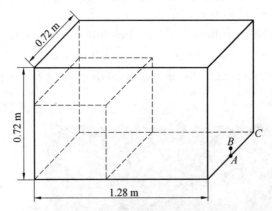

Fig. 6-5 Simulation condition of dam break

Fig. 6-6 shows the time history of the calculated pressure at point A with Case 01, Case 02, Case 04 and proposed method. Fig. 6-6 (a) compares the time history of the pressure between Case 01 and Case 02. The difference between Case 01 and Case 02 is the source term. Before 0.75 s the pressure oscillation of Case 02 is stronger than Case 01. Thus, only changing the source term causes larger pressure oscillation. Fig. 6-6 (b) shows the improvement by using the proposed source term. The pressure oscillation is dramatically reduced. Fig. 6-6 (c) shows the improvement of asymmetric gradient model. The asymmetric gradient model can further suppress the pressure oscillation.

Chapter 6 Improvement of pressure distribution in polygon wall boundary condition

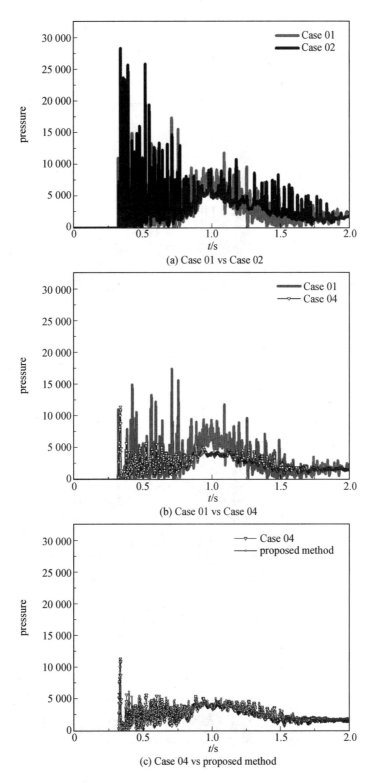

Fig. 6-6   Pressure time history in the dam break simulation at point A

Fig. 6-7 and 6-8 show the respective time history of the calculated pressure at point B and C. The pressure distribution of point B and C has the same tendency as that at point A, which illustrates that the proposed method can effectively suppress the pressure oscillation near polygons.

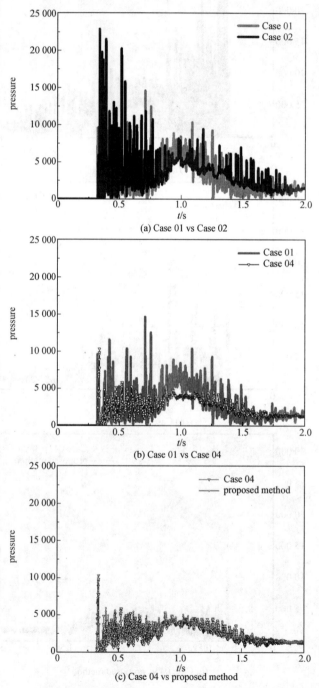

Fig. 6-7 Pressure time history in the dam break simulation at point B

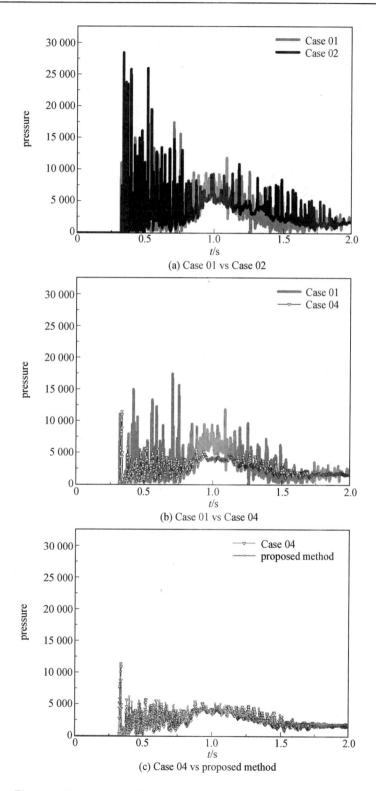

Fig. 6-8 Pressure time history in the dam break simulation at point C

The weighted average of the particle number density in Eqs. (6-20) and (6-21) is proposed to suppress the pressure oscillation of the non-uniform generated boundary particles. To compare the difference between weighted and no-weighted average of the particle number density, we assume three layers of boundary particles are arranged uniformly along the water tank as shown in Fig. 6-5. The grid points used in the BPA technique coincide with these boundary particles under initial condition. Then translate all the grid points along X, Y and Z axes simultaneously with a distance d. The dummy particles are generated at grid points and then moved to form layers of boundary particles as illustrated in Chapter 5. Using these generated boundary particles to test the time history of the pressure at reference points to show the improvement of the weighted averaged.

Fig. 6-9 shows the improvement of the weighted average of the particle number density between Case 05 and the proposed method. Fig. 6-9 (a) shows the time history of calculated pressure at point C when grid points translate $d=0.8L_0$. Fig. 6-9 (b) shows the time history of calculated pressure at point B when grid points translate $d=0.9L_0$. Other points have the same tendency as those at points B and C. In Fig. 6-9 (a), weighted average produces less pressure oscillation than without using weighted average. In Fig. 6-9 (b), the pressure oscillation without using the weighted average is stronger than that with weighted average before 1 s at point B. Although the pressure oscillation without using the weighted average is a little weaker than that with weighted average from $1\sim2$ s, sharp pressure oscillation occurs around 1.9 s without using the weighted average. On the contrary using the weighted average, the pressure gradually decreases with the time. Thus, the weighted average of the particle number density can suppress the pressure oscillation caused by the non-uniform generated boundary particles.

Fig. 6-10 shows the snapshots of the dam break simulation between Case 01 and the proposed method at $t=0.5, 1.0, 1.5, 2.0, 2.5, 3.0$ s. The color of the particles represents the pressure. The proposed method obtains better pressure distribution than Case 01.

Chapter 6 Improvement of pressure distribution in polygon wall boundary condition

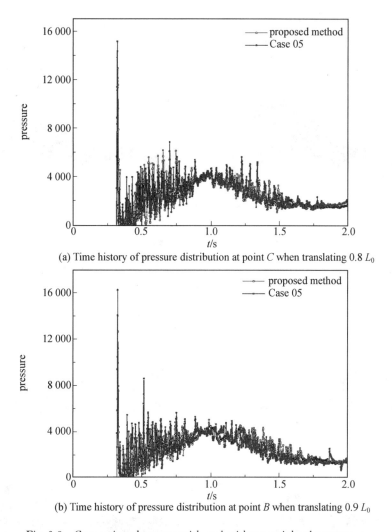

(a) Time history of pressure distribution at point C when translating $0.8\ L_0$

(b) Time history of pressure distribution at point B when translating $0.9\ L_0$

Fig. 6-9 Comparison between with and without weighted average

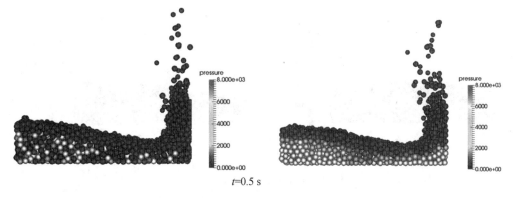

$t=0.5$ s

Fig. 6-10 Snapshots of the pressure distribution in dam break simulation

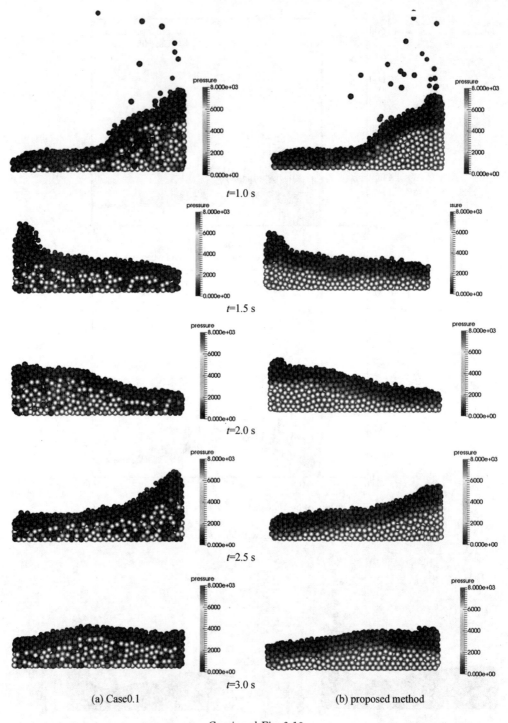

(a) Case0.1 (b) proposed method

Continued Fig. 6-10

The position change of the leading edge is compared subsequently. The simulation condition is illustrated in Fig. 6-11. The length, width and the height of the water tank are 8 m, 2 m and 8 m, respectively. The length and height of the fluid are 2 m and 4 m, respectively. The total number of fluid particles is 245 000. The diameter of the fluid particle is 0.04 m and the influence radius $r_e = 3.1L_0$.

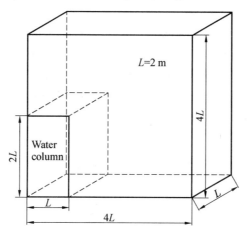

Fig. 6-11 Geometry of collapse of a water column

The comparison for the position change of the leading edge compared among Case 01, original MPS, and the proposed method is shown in Fig. 6-12. The horizontal axis indicates the non-dimensional time $t(2g/L)^{0.5}$, and the vertical axis represents the non-dimensional position of water column's edge. The simulation results compare with the experimental results by Martin and Moyce[114]. The light grey line represents the original MPS proposed by Koshizuka & Oka[40], and the black line is the result of Case 01. The proposed method is given by grey line. Case 01 has better agreement with the experimental results than original MPS from Fig. 6-12 because polygon boundary

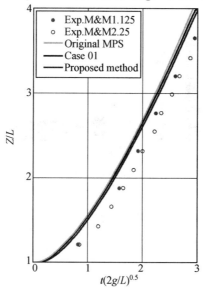

Fig. 6-12 Comparison for positon of leading edge of collapsed water column

condition can represent the wall boundary more accurately than the wall boundary composed by particles. The boundary particles used in the original MPS influence the movement of the fluid particles as illustrated in [72]. Although the result of the proposed method is slightly different from the experimental results, it is better than Case 01. Fig. 6-13 shows the snapshots of the high resolution simulation results between Case 01 and the proposed method at $t =$ 0.5, 1.0, 1.5, 2.0, 2.5, 2.7 s. The proposed method produces more stable pressure distribution than Case 01.

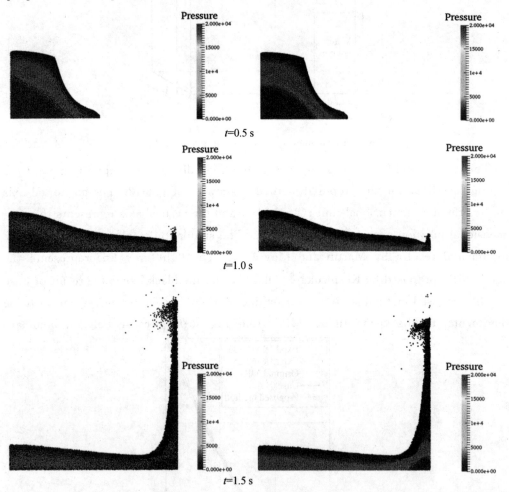

Fig. 6-13 Snapshots of the pressure distribution with high resolution

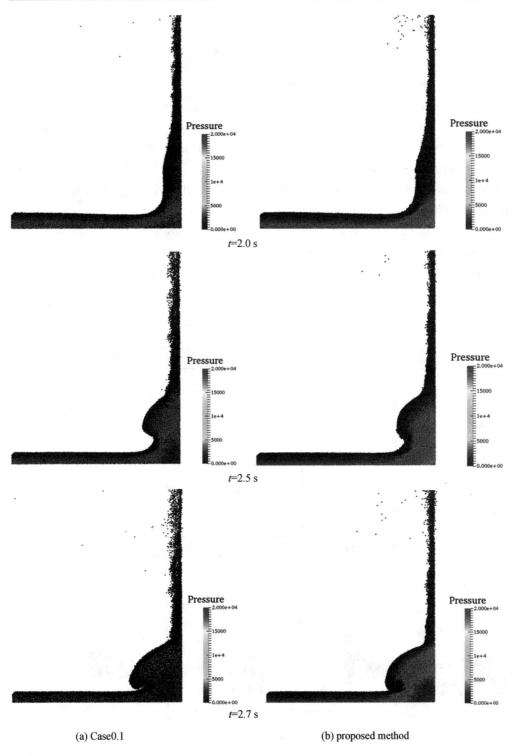

(a) Case0.1   (b) proposed method

Continued Fig. 6-13

6.3.3 Complex geometry simulations

Polygon boundary condition[41] is proposed to simulate complex geometries. To verify the improvement of the proposed method in complex geometries, two complex geometries are tested. The first example is a water tank with a wedge. The sketch of the problem is shown in Fig. 6-14. The length, width and height of a water tank are 1.68, 0.72, and 1.0 m, respectively. The length and height of the fluids are 0.68 and 0.52 m, respectively. There is a wedge inside the water tank. The left plane of the wedge is a slope and the right of the wedge is perpendicular to the bottom plane of the water tank. The number of fluid particles is 4 284. The simulation conditions are the same as section 4.2. Fig. 6-15 shows the simulation results between the proposed method and Case 01. The color represents the pressure. From Fig. 6-15, Case 01 suffers from severe pressure oscillation. On the contrary, the proposed method obtains smooth pressure distribution.

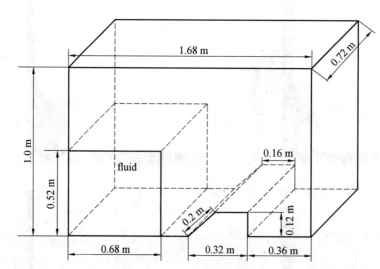

Fig. 6-14 Sketch of a water tank with a wedge

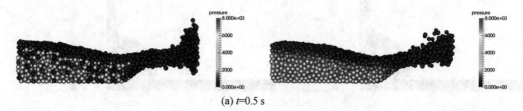

(a) $t=0.5$ s

Fig. 6-15 Snapshots of the pressure distribution

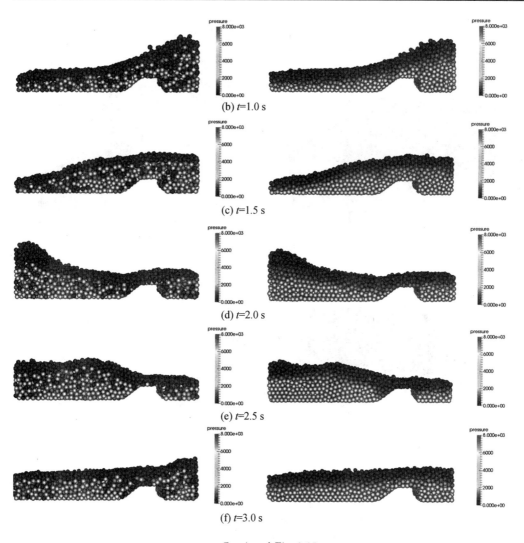

Continued Fig. 6-15

The second example is a complex geometry with curved surfaces as shown in Fig. 6-16 which is the same as Fig. 6-16. The diameter of fluid particle is $L_0 = 0.0001$ m and r_e is still $3.1L_0$. The simulation conditions are the same as above example. The fluid particles are arranged on the left side of the geometry with the initial velocity $u_x = 1, u_y = 0, u_z = 0$ m/s as shown in Fig. 6-17. Fig. 6-18 shows the snapshots of the simulation results between the proposed method and Case 01 at $t = 0.5, 1.0, 1.5, 2.0, 3.0$ s. The color represents the pressure. Since the fluid particles are stuck at the narrow mouth, the pressure increases. The pressure continuously increases until $t = 0.14$ s when the pressure reaches maximum. Then the velocity of particles becomes slowly and the pressure decreases as shown at 0.16 and 0.18 s. The simulation results indicate that the proposed method can obtain smooth pressure distribution in the simulations of complex geometries.

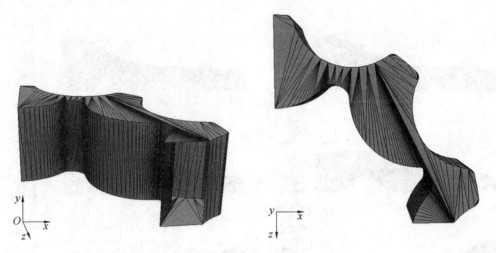

Fig. 6-16　Complex geometry composed by polygons

Fig. 6-17　Initial set-up of the fluids

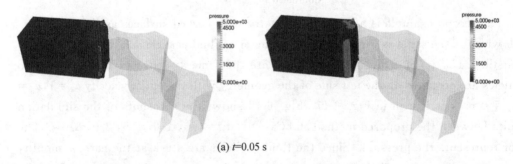

(a) $t=0.05$ s

Fig. 6-18　Snapshots of the complex geometry at $t = 0.05, 0.1, 0.14, 0.186$ s

Chapter 6 Improvement of pressure distribution in
polygon wall boundary condition • 95 •

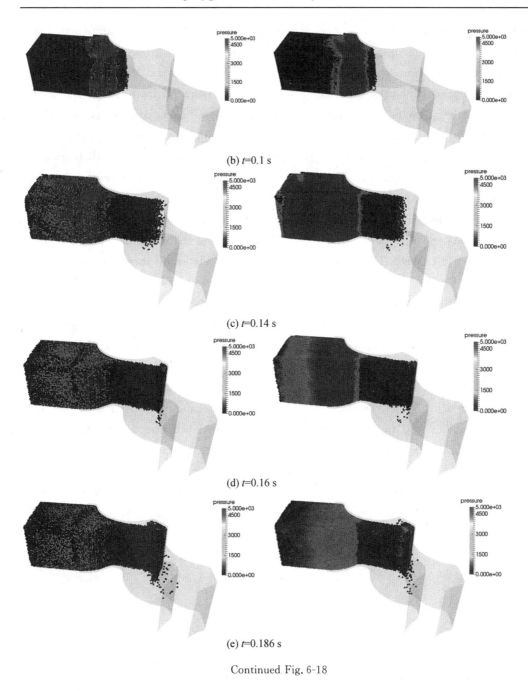

(b) *t*=0.1 s

(c) *t*=0.14 s

(d) *t*=0.16 s

(e) *t*=0.186 s

Continued Fig. 6-18

6.4 Summary

The polygon boundary condition can accurately represent the wall boundary of geometries and the complex geometries can be easily simulated. However, it suffers from severe pressure oscillations. In this study we analyzed the problems of Poisson's equation in the polygon boundary condition and pointed out that the inaccurate wall part of Poisson's equation mainly caused the numerical oscillation of pressure. To improve the pressure distribution, the source term proposed by Tanaka and Masunaga[104] is introduced to suppress the pressure oscillation of fluid particles far from the polygons. A proportion factor of the particle number density is proposed to accurately represent the contribution of the fluid and wall parts of the Poisson's equation when fluid particles are close to the polygons. Thus, the pressure oscillation near the polygon wall boundary can be effectively suppressed. To suppress the pressure oscillation caused by the generated boundary particles, weighted average of the particle number density is introduced into the source term. By doing so, the inaccurate wall part of the Poisson's equation can be improved for the first time. The pressure distribution near polygons can be dramatically improved. Then the asymmetric form of the gradient model proposed by Khayyer and Gotoh[107] was adopted to further improve the pressure calculation especially near the polygons and suppress the pressure oscillation. Since the distance from the fluid particles to the polygons can be simply calculated by interpolation, the wall weight function does not need to calculate if the distance is larger than effective radius. thus, Thus, the proposed method can still retain the high efficiency.

To illustrate the improvement of the proposed method in polygon boundary condition, four three dimensional exampleswere verified against other models. The hydrostatic simulation illustrated that the proposed method can effectively improve the pressure calculation especially near polygons. The dam break simulation showed the improvement of the proposed method to reduce the pressure oscillation. Simulations of two complex geometries demonstrated that the proposed method can be easily generalized to the complex geometries without special treatment to the polygon wall boundary.

References

[1] CRESPO A J C. Application of the smoothed particle hydrodynamics model SPHysics to free surface hydrodynamics [D]. Vigo: Ph. D. thesis, University of De Vigo, 2008.

[2] LIU G R, LIU M B. Smoothed particle hydrodynamics: a meshfree particle method [M]. Singapore: World Scientific, 2003.

[3] GODUNOV S K, ZABRODIN A V, PROKOPOV G P. A computational scheme for two-dimensional nonstationary problems of gas dynamics and calculation of the flow from a shock wave approaching a stationary state[J]. Zhurnal Vychislitel'noi Matematiki i Matematicheskoi Fiziki, 1961, 1(6): 1020-1050.

[4] CHANG C H, LIOU M S. A robust and accurate approach to computing compressible multiphase flow: stratified flow model and $AUSM^+$-up scheme[J]. Journal of Computational Physics, 2007, 225(1): 840-873.

[5] RICHTMYER R D. A survey of difference methods for non-steady fluid dynamics [M]. Boulder, Colorado: National Center for Atmospheric Research, 1963.

[6] ZIENKIEWICZ O C, TAYLOR R L, ZIENKIEWICZ O C, et al. The finite element method[M]. London: McGraw-hill, 1977.

[7] DHATT G, LEFRANĀ E, TOUZOT G. Finite element method[M]. Hoboken, New Jersey: John Wiley & Sons, 2012.

[8] VERSTEEG H K, MALALASEKERA W. An introduction to computational fluid dynamics: the finite volume method[M]. New York: Pearson Education, 2007.

[9] PESKIN C S. The fluid dynamics of heart valves: experimental, theoretical, and computational methods[J]. Annual Review of Fluid Mechanics, 1982, 14(1): 235-259.

[10] FULLSACK P. An arbitrary Lagrangian-Eulerian formulation for creeping flows and its application in tectonic models[J]. Geophysical Journal International, 1995, 120(1): 1-23.

[11] MORESI L, DUFOUR F, MÜHLHAUS H B. A Lagrangian integration point

finite element method for large deformation modeling of viscoelastic geomaterials [J]. Journal of Computational Physics, 2003, 184(2): 476-497.

[12] MORESI L, MÜHLHAUS H B, DUFOUR F. Particle-in-cell solutions for creeping viscous flows with internal interfaces[J]. In Bifurcation and Localization in Soils and Rocks, 2001(2): 345-353.

[13] HARLOW F H, WELCH J E. Numerical calculation of time-dependent viscous incompressible flow of fluid with free surface [J]. The physics of fluids, 1965, 8(12): 2182-2189.

[14] SULSKY D, SCHREYER H L. Axisymmetric form of the material point method with applications to upsetting and Taylor impact problems[J]. Computer Methods in Applied Mechanics and Engineering, 1996, 139(1-4): 409-429.

[15] ONATE E, IDELSOHN S, ZIENKIEWICZ O, et al. A finite point method in computational mechanics. Applications to convective transport and fluid flow[J]. International Journal for Numerical Methods in Engineering, 1996,39: 3839-3866.

[16] ONATE E, IDELSOHN S, ZIENKIEWICZ O C, et al. A stabilized finite point method for analysis of fluid mechanics problems [J]. Computer Methods in Applied Mechanics and Engineering, 1996, 139(1-4): 315-346.

[17] ONATE E, IDELSOHN S. A mesh-free finite point method for advective-diffusive transport and fluid flow problems[J]. Computational Mechanics, 1998, 21(4-5): 283-292.

[18] OÑATE E, PERAZZO F, MIQUEL J. A finite point method for elasticity problems[J]. Computers & Structures, 2001, 79(22-25): 2151-2163.

[19] NAYROLES B, TOUZOT G, VILLON P. Generalizing the finite element method: diffuse approximation and diffuse elements [J]. Computational Mechanics, 1992,10: 307-318.

[20] BELYTSCHKO T, LU Y Y, GU L. Element-free Galerkin methods [J]. International Journal for Numerical Methods in Engineering, 1994, 37 (2): 229-256.

[21] LU Y Y, BELYTSCHKO T, GU L. A new implementation of the element free Galerkin method[J]. Computer Methods in Applied Mechanics and Engineering, 1994, 113(3-4): 397-414.

[22] BELYTSCHKO T, GU L, LU Y Y. Fracture and crack growth by element free

Galerkin methods [J]. Modelling and Simulation in Materials Science and Engineering, 1994, 2(3A): 519.

[23] BELYTSCHKO T, LU Y Y, GU L, et al. Element-free Galerkin methods for static and dynamic fracture[J]. International Journal of Solids and Structures, 1995, 32(17-18): 2547-2570.

[24] LIU W K, JUN S, ZHANG Y F. Reproducing kernel particle methods[J]. International Journal for Numerical Methods in Fluids, 1995, 20(8-9): 1081-1106.

[25] LIU W K, JUN S, LI S, et al. Reproducing kernel particle methods for structural dynamics[J]. International Journal for Numerical Methods in Engineering, 1995, 38(10): 1655-1679.

[26] CHEN J S, PAN C, WU C T, et al. Reproducing kernel particle methods for large deformation analysis of non-linear structures[J]. Computer Methods in Applied Mechanics and Engineering, 1996, 139(1-4): 195-227.

[27] LIU W K, LI S, BELYTSCHKO T. Moving least-square reproducing kernel methods (I) methodology and convergence[J]. Computer Methods in Applied Mechanics and Engineering, 1997, 143(1-2): 113-154.

[28] DUARTE C A, ODEN J T. An hp adaptive method using clouds[J]. Computer Methods in Applied Mechanics and Engineering, 1996, 139(1-4): 237-262.

[29] DUARTE C A, ODEN J T. Hp clouds-an hp meshless method[J]. Numerical Methods for Partial Differential Equations, 1996, 12(6): 673-706.

[30] LISZKA T J, DUARTE C A, TWORZYDLO W W. Hp-meshless cloud method [J]. Computer Methods in Applied Mechanics and Engineering, 1996, 139(1-4): 263-288.

[31] ATLURI S N, ZHU T. A new meshless local Petrov-Galerkin (MLPG) approach in computational mechanics[J]. Computational Mechanics, 1998, 22(2): 117-127.

[32] ATLURI S N, KIM H G, CHO J Y. A critical assessment of the truly meshless local Petrov-Galerkin (MLPG), and local boundary integral equation (LBIE) methods[J]. Computational Mechanics, 1999, 24(5): 348-372.

[33] ATLURI S N, ZHU T. The meshless local Petrov-Galerkin (MLPG) approach for solving problems in elasto-statics[J]. Computational Mechanics, 2000, 25(2-3): 169-179.

[34] ATLURI S N, ZHU T. New concepts in meshless methods[J]. International

Journal for Numerical Methods in Engineering, 2000, 47(1-3): 537-556.

[35] LIU G R, GU Y T. A point interpolation method for two-dimensional solids[J]. International Journal for Numerical Methods in Engineering, 2001, 50 (4): 937-951.

[36] LIU G R, YAN L, WANG J G, et al. Point interpolation method based on local residual formulation using radial basis functions[J]. Structural Engineering and Mechanics, 2002, 14(6): 713-732.

[37] LIMA N Z, MESQUITA R C. Point interpolation methods based on weakened-weak formulations [J]. Journal of Microwaves, Optoelectronics and Electromagnetic Applications, 2013, 12(2): 506-523.

[38] GINGOLD R A, MONAGHAN J J. Smoothed particle hydrodynamics: theory and application to non-spherical stars [J]. Monthly Notices of the Royal Astronomical Society, 1977, 181(3): 375-389.

[39] LUCY L B. A numerical approach to the testing of the fission hypothesis[J]. The Astronomical Journal, 1977, 82: 1013-1024.

[40] KOSHIZUKA S, OKA Y. Moving-particle semi-implicit method for fragmentation of incompressible fluid[J]. Nuclear Science and Engineering, 1996, 123(3): 421-434.

[41] HARADA T. Improvement of wall boundary calculation model for MPS method [J]. Transactions of JSCES,2008(20080006).

[42] XIE J, TAI Y C, JIN Y C. Study of the free surface flow of water-kaolinite mixture by moving particle semi-implicit (MPS) method[J]. International Journal for Numerical and Analytical Methods in Geomechanics, 2014, 38(8): 811-827.

[43] GOTOH H, SAKAI T. Key issues in the particle method for computation of wave breaking[J]. Coastal Engineering, 2006, 53(2-3): 171-179.

[44] SHIBATA K, KOSHIZUKA S. Numerical analysis of shipping water impact on a deck using a particle method[J]. Ocean Engineering, 2007, 34(3-4): 585-593.

[45] SHAKIBAEINIA A, JIN Y C. MPS mesh-free particle method for multiphase flows[J]. Computer Methods in Applied Mechanics and Engineering, 2012, 229: 13-26.

[46] KOSHIZUKA S, IKEDA H, OKA Y. Numerical analysis of fragmentation mechanisms in vapor explosions [J]. Nuclear Engineering and Design,

1999, 189(1-3): 423-433.

[47] SUN X, SAKAI M, SAKAI M T, et al. A Lagrangian-Lagrangian coupled method for three-dimensional solid-liquid flows involving free surfaces in a rotating cylindrical tank[J]. Chemical Engineering Journal, 2014, 246: 122-141.

[48] SUN Z, XI G, CHEN X. A numerical study of stir mixing of liquids with particle method[J]. Chemical Engineering Science, 2009, 64(2): 341-350.

[49] ATAIE-ASHTIANI B, FARHADI L. A stable moving-particle semi-implicit method for free surface flows[J]. Fluid Dynamics Research, 2006, 38(4): 241-256.

[50] SHAKIBAEINIA A, JIN Y C. A weakly compressible MPS method for modeling of open-boundary free-surface flow[J]. International Journal for Numerical Methods in Fluids, 2010, 63(10): 1208-1232.

[51] BELYTSCHKO T, KRONGAUZ Y, ORGAN D, et al. Meshless methods: an overview and recent developments[J]. Computer Methods in Applied Mechanics and Engineering, 1996, 139(1-4): 3-47.

[52] KOSHIZUKA S. A particle method for incompressible viscous flow with fluid fragmentation[J]. Comput. Fluid Dyn. J., 1995, 4: 29-46.

[53] SHAO S, LO E Y M. Incompressible SPH method for simulating Newtonian and non-Newtonian flows with a free surface[J]. Advances in Water Resources, 2003, 26(7): 787-800.

[54] KON T, NATSUI S, UEDA S, et al. Analysis of effect of packed bed structure on liquid flow in packed bed using moving particle semi-implicit method[J]. ISIJ International, 2015, 55(6): 1284-1290.

[55] MUSTARI A P A, OKA Y, FURUYA M, et al. 3D simulation of eutectic interaction of Pb-Sn system using moving particle semi-implicit (MPS) method [J]. Annals of Nuclear Energy, 2015, 81: 26-33.

[56] TAYEBI A, JIN Y. Development of moving particle explicit (MPE) method for incompressible flows[J]. Computers & Fluids, 2015, 117: 1-10.

[57] FEDERICO I, MARRONE S, COLAGROSSI A, et al. Simulating 2D open-channel flows through a SPH model[J]. European Journal of Mechanics-B/Fluids, 2012, 34: 35-46.

[58] DALRYMPLE R A, ROGERS B D. Numerical modeling of water waves with the

SPH method[J]. Coastal Engineering, 2006, 53(2-3): 141-147.

[59] MONAGHAN J J. Simulating free surface flows with SPH[J]. Journal of Computational Physics, 1994, 110(2): 399-406.

[60] BONET J, LOK T S L. Variational and momentum preservation aspects of smooth particle hydrodynamic formulations[J]. Computer Methods in Applied Mechanics and Engineering, 1999, 180(1-2): 97-115.

[61] VIOLEAU D, ISSA R. Numerical modelling of complex turbulent free-surface flows with the SPH method: an overview[J]. International Journal for Numerical Methods in Fluids, 2007, 53(2): 277-304.

[62] MORRIS J P, FOX P J, ZHU Y. Modeling low Reynolds number incompressible flows using SPH[J]. Journal of Computational Physics, 1997, 136(1): 214-226.

[63] TAKEDA H, MIYAMA S M, SEKIYA M. Numerical simulation of viscous flow by smoothed particle hydrodynamics[J]. Progress of Theoretical Physics, 1994, 92(5): 939-960.

[64] JONES V, YANG Q, MCCUE-WEIL L. SPH boundary deficiency correction for improved boundary conditions at deformable surfaces[J]. Ciencia y tecnología de buques, 2010, 4(7): 21-30.

[65] COLAGROSSI A, LANDRINI M. Numerical simulation of interfacial flows by smoothed particle hydrodynamics [J]. Journal of Computational Physics, 2003, 191(2): 448-475.

[66] ADAMI S, HU X Y, ADAMS N A. A generalized wall boundary condition for smoothed particle hydrodynamics[J]. Journal of Computational Physics, 2012, 231(21): 7057-7075.

[67] MONAGHAN J J. Smoothed particle hydrodynamics[J]. Annual Review of Astronomy and Astrophysics, 1992, 30(1): 543-574.

[68] BECKER M, TESCHNER M. Weakly compressible SPH for free surface flows [C]. In Proceedings of the 2007 ACM SIGGRAPH/Eurographics Symposium on Computer Animation, Eurographics Association, 2007:209-217.

[69] BECKER M, TESSENDORF H, TESCHNER M. Direct forcing for Lagrangian rigid-fluid coupling [J]. IEEE Transactions on Visualization and Computer Graphics, 2009, 15(3): 493-503.

[70] SUN Z, LIANG Y, XI G. Numerical simulation of the flow in straight blade

agitator with the MPS method[J]. International Journal for Numerical Methods in Fluids, 2011, 67(12): 1960-1972.

[71] GAMBARUTO A M. Computational haemodynamics of small vessels using the moving particle semi-implicit (MPS) method [J]. Journal of Computational Physics, 2015, 302: 68-96.

[72] LI D, SUN Z, CHEN X, et al. Analysis of wall boundary in moving particle semi-implicit method and a novel model of fluid-wall interaction [J]. International Journal of Computational Fluid Dynamics, 2015, 29(3-5): 199-214.

[73] OGER G, LEROY C, JACQUIN E, et al. Alessandrini, specific pre/post treatments for 3-D SPH applications through massive HPC simulations [C]. In Proc. 4th International SPHERIC workshop, 2009: 27-29.

[74] CUMMINS S J, RUDMAN M. A SPH projection method [J]. Journal of Computational Physics, 1999, 152(2): 584-607.

[75] POZORSKI J, WAWREŃCZUK A. SPH computation of incompressible viscous flows[J]. Journal of Theoretical and Applied Mechanics, 2002, 40(4): 917-937.

[76] AKIMOTO H. Numerical simulation of the flow around a planing body by MPS method[J]. Ocean Engineering, 2013, 64: 72-79.

[77] CHERFILS J M, PINON G, RIVOALEN E. Josephine: a parallel SPH code for free-surface flows [J]. Computer Physics Communications, 2012, 183 (7): 1468-1480.

[78] YILDIZ M, ROOK R A, SULEMAN A. SPH with the multiple boundary tangent method [J]. International Journal for Numerical Methods in Engineering, 2009, 77(10): 1416-1438.

[79] MARRONE S, ANTUONO M, COLAGROSSI A, et al. δ-SPH model for simulating violent impact flows[J]. Computer Methods in Applied Mechanics and Engineering, 2011, 200(13-16): 1526-1542.

[80] FRIES T P, MATTHIES H G. Classification and overview of meshfree methods [D]. Braunschweig: Department of Mathematics and Computer Science, Technical University of Braunschweig, 2003.

[81] PARK S, JEUN G. Coupling of rigid body dynamics and moving particle semi-implicit method for simulating isothermal multi-phase fluid interactions [J]. Computer Methods in Applied Mechanics and Engineering, 2011, 200 (1-4):

130-140.

[82] LEE B H, PARK J C, KIM M H, et al. Step-by-step improvement of MPS method in simulating violent free-surface motions and impact-loads[J]. Computer Methods in Applied Mechanics and Engineering, 2011, 200(9-12): 1113-1125.

[83] MONAGHAN J J, KOS A. Solitary waves on a Cretan beach[J]. Journal of Waterway, Port, Coastal, and Ocean Engineering, 1999, 125(3): 145-155.

[84] MONAGHAN J J. Simulating gravity currents with {SPH}{III}{B}oundary forces[J]. Mathematics Reports and Preprints, 1995, 2: 17-33.

[85] MONAGHAN J J. Improved modelling of boundaries[J]. SPH Tech. Note, 1995, 95:2-30.

[86] DE LEFFE M, LE TOUZÉ D, ALESSANDRINI B. Normal flux method at the boundary for SPH[C]. In 4th Int. SPHERIC Workshop (SPHERIC 2009).

[87] MONAGHAN J J, KAJTAR J B. SPH particle boundary forces for arbitrary boundaries[J]. Computer Physics Communications, 2009, 180(10): 1811-1820.

[88] MACIÀ F, GONZÁLEZ L M, CERCOS-PITA J L, et al. A boundary integral SPH formulation: consistency and applications to ISPH and WCSPH[J]. Progress of Theoretical Physics, 2012, 128(3): 439-462.

[89] FERRAND M, LAURENCE D R, ROGERS B D, et al. Unified semi-analytical wall boundary conditions for inviscid, laminar or turbulent flows in the meshless SPH method[J]. International Journal for Numerical Methods in Fluids, 2013, 71(4): 446-472.

[90] SHAO S. Incompressible SPH simulation of wave breaking and overtopping with turbulence modelling[J]. International Journal for Numerical Methods in Fluids, 2006, 50(5): 597-621.

[91] HU X Y, ADAMS N A. An incompressible multi-phase SPH method[J]. Journal of Computational Physics, 2007, 227(1): 264-278.

[92] LEROY A, VIOLEAU D, FERRAND M, et al. Unified semi-analytical wall boundary conditions applied to 2-D incompressible SPH [J]. Journal of Computational Physics, 2014, 261: 106-129.

[93] MAYRHOFER A, FERRAND M, KASSIOTIS C, et al. Unified semi-analytical wall boundary conditions in SPH: analytical extension to 3-D[J]. Numerical Algorithms, 2015, 68(1): 15-34.

[94] MAYRHOFER A, FERRAND M, KASSIOTIS C, et al. Unified semi-analytical wall boundary conditions in SPH: analytical extension to 3-D[J]. Numerical Algorithms, 2015, 68(1): 15-34.

[95] KUASEGARAM S, BONET J, LEWIS R W, et al. A variational formulation based contact algorithm for rigid boundaries in two-dimensional SPH applications [J]. Computational Mechanics, 2004, 33(4): 316-325.

[96] FELDMAN J, BONET J. Dynamic refinement and boundary contact forces in SPH with applications in fluid flow problems[J]. International Journal for Numerical Methods in Engineering, 2007,72: 295-324.

[97] 渡辺高志，桝谷浩，三橋祐太. 壁面境界の大変形を考慮した粒子法の計算手法に関する基礎的研究[C]. 日本計算工学会論文集，2013 (2013): 20130021-20130021.

[98] MEYER M, DESBRUN M, SCHRÖDER P, et al. Discrete differential-geometry operators for triangulated 2-manifolds[M]//Visualization and Mathematics Ⅲ. Springer, Berlin, Heidelberg, 2003: 35-57.

[99] YAMADA Y, SAKAI M, MIZUTANI S, et al. Numerical simulation of three-dimensional free-surface flows with explicit moving particle simulation method[J]. Transactions of the Atomic Energy Society of Japan, 2011, 10(3): 185-193.

[100] MITSUME N, YOSHIMURA S, MUROTANI K, et al. Explicitly represented polygon wall boundary model for the explicit MPS method[J]. Computational Particle Mechanics, 2015, 2(1): 73-89.

[101] SHIBATA K, KOSHIZUKA S, OKA Y. Numerical analysis of jet breakup behavior using particle method[J]. Journal of Nuclear Science and Technology, 2004, 41(7): 715-722.

[102] IKEDA H. Numerical analysis of fragmentation processes in vapor explosions using particle method[D]. Tokyo:PhD thesis, The University of Tokyo, 1999.

[103] HIBI S. A study on reduction of unusual pressure fluctuation of MPS method[J]. J Kansai Soc NA Jpn, 2004, 241: 125-131.

[104] TANAKA M, MASUNAGA T. Stabilization and smoothing of pressure on MPS method by quasi-compressibility[J]. Transactions of JSCES, 2008(20080025).

[105] KOH C G, LUO M, GAO M, et al. Modelling of liquid sloshing with constrained floating baffle[J]. Computers & Structures, 2013, 122: 270-279.

[106] KOH C G, GAO M, LUO C. A new particle method for simulation of incompressible free surface flow problems [J]. International Journal for Numerical Methods in Engineering, 2012, 89(12): 1582-1604.

[107] KHAYYER A, GOTOH H. Development of CMPS method for accurate water-surface tracking in breaking waves [J]. Coastal Engineering Journal, 2008, 50(02): 179-207.

[108] KONDO M, KOSHIZUKA S. Improvement of stability in moving particle semi-implicit method[J]. International Journal for Numerical Methods in Fluids, 2011, 65(6): 638-654.

[109] KHAYYER A, GOTOH H. Modified moving particle semi-implicit methods for the prediction of 2D wave impact pressure [J]. Coastal Engineering, 2009, 56(4): 419-440.

[110] MOLTENI D, COLAGROSSI A. A simple procedure to improve the pressure evaluation in hydrodynamic context using the SPH [J]. Computer Physics Communications, 2009, 180(6): 861-872.

[111] HOSSEINI S M, FENG J J. Pressure boundary conditions for computing incompressible flows with SPH[J]. Journal of Computational physics, 2011, 230(19): 7473-7487.

[112] XU R, STANSBY P, LAURENCE D. Accuracy and stability in incompressible SPH (ISPH) based on the projection method and a new approach[J]. Journal of Computational Physics, 2009, 228(18): 6703-6725.

[113] LIND S J, XU R, STANSBY P K, et al. Incompressible smoothed particle hydrodynamics for free-surface flows: a generalised diffusion-based algorithm for stability and validations for impulsive flows and propagating waves[J]. Journal of Computational Physics, 2012, 231(4): 1499-1523.

[114] MARTIN J C, MOYCE W J, PENNEY W G, et al. Part IV. An experimental study of the collapse of liquid columns on a rigid horizontal plane[J]. Phil. Trans. R. Soc. Lond. A, 1952, 244(882): 312-324.

Noun index

B

Boundary forces 3.1

Boundary particle arrangement technique 5.1

C

Collision coefficient 5.2

Construction of boundary particle 5.1

Cylindrical coordinate system 3.1

D

Diffuse element method 1.1

Dirichlet boundary condition 2.4

Discretization of curvature 3.2

Dummy particles 3.1

E

Element free Galerkin method 1.1

Eulerian method 1.1

Explicit polygon wall boundary condition 3.1

F

Finite difference method 1.1

Finite element method 1.1

Finite point method 1.1

Finite volume method 1.1

G

Gradient term of polygon wall boundary condition 3.3

H

HP-cloud method 1.1

I

Incompressible condition 2.2

Inlet boundary condition 3.1

L

Lagrangian methods 1.1

Laplacian model 2.2

Laplacian model coefficient 2.2

Linear momentum conservative gradient model 6.2

M

Mass conservation equation 2.1

Meshless local Petrov-Galerkin method 1.1

Mirror particles 3.1

Moving least-squares kernel 3.1

Moving particle semi-implicit method 1.1, 2.1, 3.2

Multiply boundary tangent method 3.1

N

Navier-Stokes equation 2.1

Normal vector of grid point 3.2

No-slip boundary condition 3.1

O

Outlet boundary condition 3.1

P

Particle number density polygon wall boundary condition 3.3

Point interpolation method 1.1, 3.1

Poisson equation 2.3

Polygon wall boundary condition 1.2

Pressure gradient model 2.2

R

Reproduced kernel particle method 1.1

S

Smoothed particle hydrodynamics 1.1

Solid wall boundary condition   2.4

Source term of polygon wall boundary condition 3.3

Source term of pressure 2.2

Surface detection 6.2

U

Unified semi-analytical wall boundary condition 3.1

V

Viscosity term 2.2

Viscosity term of polygon wall boundary condition 3.3

W

Wall weight function   4.2

Weight function 2.1

Weighted average velocity 5.3